AUSTRALIA
IN THE
WORLD

An Introduction to Australian Foreign Policy

For Jan and Gudie

AUSTRALIA IN THE WORLD

An Introduction to Australian Foreign Policy

Gary Smith

Dave Cox

Scott Burchill

Melbourne

OXFORD UNIVERSITY PRESS

Oxford   Auckland   New York

OXFORD UNIVERSITY PRESS AUSTRALIA

Oxford New York
Athens Auckland Bangkok Bombay
Calcutta Cape Town Dar es Salaam Delhi
Florence Hong Kong Istanbul Karachi
Kuala Lumpur Madras Madrid Melbourne
Mexico City Nairobi Paris Port Moresby
Singapore Taipei Tokyo Toronto

and associated companies in
Berlin Ibadan

OXFORD is a trade mark of Oxford University Press

National Library of Australia
Cataloguing-in-Publication data:

Smith, Gary (Gary Kennedy).
Australia in the world: an introduction to Australian foreign policy.

Bibliograhy.
  Includes index.
  ISBN 0 19 553477 8.

  1. Australia — Foreign relations. 2. Australia —
  Politics and government — 20th
  century. I Cox, Dave, 1956– . II. Burchill, Scott. III.
  Title.

327.94

Edited by Lucy Davison
Typeset by Desktop Concepts P/L, Melbourne
Printed through Bookpac Production Services, Singapore
Published by Oxford University Press,
253 Normanby Road, South Melbourne, Australia

Contents

Abbreviations

ACTU	Australian Council of Trade Unions
AFTA	ASEAN Free Trade Area
AGPS	Australian Government Publishing Service
ALP	Australian Labor Party
AMIC	Australian Mining Industry Council
ANU	Australian National University
ANZAM	Australia, New Zealand and Malaya
ANZUS	Australia, New Zealand and the United States
APEC	Asia-Pacific Economic Cooperation
ARF	ASEAN Regional Forum
ASEAN	Association of South-East Asian Nations
CD	Conference on Disarmament
COP-1	Conference of the Parties to the FCCC
CSCE	Conference on Security and Cooperation in Europe
CTBT	Comprehensive Test Ban Treaty
CWC	Chemical Weapons Convention
DFAT	Department of Foreign Affairs and Trade
DLP	Democratic Labor Party
EAEC	East Asian Economic Caucus
EC	European Community
EEC	European Economic Community
EEP	Export Enhancement Program (United States)
EPAC	Economic, Planning and Advisory Committee
EU	European Union
FCCC	Framework Convention on Climate Change (UN)
FIRB	Foreign Investment Review Board
GATT	General Agreement on Tariffs and Trade
GDP	gross domestic product
GIC	good international citizenship
ICBM	intercontinental ballistic missile
IPCC	Intergovernmental Panel on Climatic Change

IPE	international political economy
JUSCANZ	Japan, the United States, Canada, Australia and New Zealand
MAD	mutual assured destruction
MP	Member of Parliament
NAFTA	New Zealand and Australia Free Trade Agreement
NAM	Non-Aligned Movement
NATO	North Atlantic Treaty Organisation
NDP	Nuclear Disarmament Party
NPT	(Nuclear) Non-Proliferation Treaty
OECD	Organization for Economic Cooperation and Development
OPM	Organisasi Papua Merdeka (Free Papua Movement)
PM&C	Department of the Prime Minister and Cabinet
PRC	People's Republic of China
RSPacS	Research School of Pacific Studies (ANU)
SEATO	South-East Asia Treaty Organisation
START	Strategic Arms Reduction Treaty
TNC	transnational corporation
UN	United Nations
UNCSD	United Nations Commission on Sustainable Development
UNEF	United Nations Emergency Force
UNEP	United Nations Environment Programme
UNMOGIP	United Nations Military Observation Group in India and Pakistan
UNTAC	United Nations Transitional Authority in Cambodia
USSR	Union of Soviet Socialist Republics
WMO	World Meteorological Organization
WTO	World Trade Organization

Introduction

Australian foreign policy at the end of the century must deal with a rapidly changing world. Australia can no longer rely on the cold war framework that, for forty years, united it with great power allies through a commitment to a common cause. The shifting political, military, and economic parameters of international relations challenge Australia to define and pursue its political, military, and economic interests and relationships in new and more complex ways. *Australia in the World* presents a comprehensive introduction to Australian foreign policy before, during, and after the cold war, with a major focus on the 1990s.

A crisis of intellectual confidence pervades much of the analysis of global politics, arising from the almost universal failure of diplomats and scholars to predict, or even allow themselves to imagine, the end of the cold war, the collapse of the Soviet empire, and the disintegration of the USSR. Rather than reaching for new certainties or a grand synthesis, this book seeks to identify the assumptions made by governments in their analysis of international relations, to give more attention to competing perspectives on the realities of global politics, and to explore the multifaceted and contradictory responses to change.

Australia has responded to the end of the cold war with a diplomatic energy that appears to belie any notion of intellectual crisis. Australia's foreign minister from 1987 to 1996, Senator Gareth Evans, has written, with academic and adviser Bruce Grant, a contemporary official history and justification of Australian foreign policy, *Australia's Foreign Relations: In the World of the 1990s* (1995). After thirteen years of Labor government, this is a primary source for students of Australia's foreign relations, but it presents some obvious and less obvious pitfalls. A foreign minister's history avoids, in the interest of presenting a picture of overall policy coherence, a political analysis of the 'push and shove' through which particular proposals became government policy. We are not told, for example, that the Prime Minister failed to inform the Foreign Minister that he was reversing Australia's policy on the Antarctic Minerals Convention in response to the electoral strength of environmentalists, and that the Foreign Minister had to abandon overnight all his well-rehearsed arguments about the benefits of a convention over a 'world park'. External pressures on Australian decisions are rarely discussed or admitted, and domestic dissent in Australia on foreign-policy issues is largely ignored.

Understandably, a foreign minister presents Australia's foreign-policy approach as the best of all possible approaches: that Australia is where it ought to be and where it is able to be. This perspective overplays the solutions that Australia has achieved, but it also underplays the significance of the intellectual and political struggle in which the reconstruction of Australian foreign policy is involved. Australian foreign-policy-makers are engaged in a restless search for a framework that positions Australia in the world and enables the identification and prioritisation of foreign-policy objectives for Australia's eighteen million citizens. The ANZUS alliance, once at the forefront of foreign policy, has moved from centre stage. Its importance has been eroded by the idea that Australia is part of the region and must give priority to regional concerns in its foreign policy. This new direction in Australian foreign policy is driven by economic imperatives, and by the idea that Australia is a middle power operating in multilateral coalitions and international institutions in relation to global issues vital to Australia's longer term interests. The response to the end of the cold war echoes Australia's response to the end of the Second World War, when great international change and uncertainty, and a failure of realism (reliance on the British), allowed the Labor Foreign Minister, Herbert 'Doc' Evatt, to explore rationalist multilateral agendas. The restlessness of the last decade arises from the pressure felt by governments to operate all the available levers in a time of massive change, from the difficulty any one of these ideas has in establishing itself as 'the cornerstone' of foreign policy once provided by the United States alliance, and also from the internal contradictions and competing policy prescriptions of these ideas. How much value should be placed in the residual United States alliance now that it is no longer bound together by the glue of anti-communism? And how far should Australia court and accommodate the USA? What, in the end, is the boundary of the region of which Australia is 'a part': South-East Asia? East Asia? the Asia Pacific? Are neo-liberal economic policies producing solutions to the problem of Australia's economic security, and if so, at what cost to the economic security of groups in Australia with little market power? Do middle powers have any common interests and basis for joint action — an intrinsic rationalist agenda that goes beyond momentary coalitions of expedience?

In seeking to provide answers to these questions, *Australia in the World* is organised into three parts.

Part 1, 'International Politics and Australian Foreign Policy', surveys the external and internal forces that shape Australian foreign policy. What are the post-cold war realities that matter most to Australia, and how do they affect the way Australia defines its interests? Different assumptions and perspectives — realist, rationalist, global community, international political economy (IPE) — suggest different answers to questions about the nature and significance of military security, economic security, environmental, and human rights issues. Foreign policy, like other public policy, is the outcome of a political process in which certain positions and views are adopted over rival and competing views, and are declared to be the 'national interest' (for the time being). Decisions may reflect the involvement of the prime minister or foreign minister, the bureaucratic line of departments, external pressure, or at times, a direct response to interest-group activity or wider public debate. Foreign-policy decisions may reflect broader political party traditions and may become election issues, but these influences seem much less significant with the growth of bipartisanship between the Australian Labor Party (ALP) and the Liberal–National Coalition.

Part II, 'Identities and Alignments', considers three major attempts at providing an overall framework for Australian foreign policy. As an ANZUS ally, Australia sought, for a historical moment, to make its relationship with the USA the cornerstone of foreign policy and to subordinate foreign policy to the principal objective of sustaining the alliance. The decline of the United States alliance can be linked to the rise of the assumption that 'our future lies in the region', and with assertions by prime ministers and foreign ministers that regional integration and enmeshment are the centrepiece of Australian foreign policy. Alternatively, we can see a more general shift in foreign policy from alliance bilateralism to middle-power multilateralism, based on multiple relationships and issue-oriented coalitions, and the development and utilisation of international organisation. Instead of offering a single historical narrative on Australian foreign policy, the evolving logic and politics of each of these three discourses on foreign policy — alliance, region, middle power — are presented as three interrelated historical perspectives on Australia's foreign relations.

Part III, 'Issues in Australia's Foreign Relations', moves from attempts to develop large organising concepts for Australian foreign policy to a consideration of five major issues, each with its own dynamics and history: global military security, regional military security, economic security, environment, and human rights. In each case, the scope of the issue is outlined, and Australia's assumptions, objectives, actions, and responses are presented, analysed, and evaluated. The restlessness in Australian foreign policy is reflected in the restlessness of these analyses, which have a bias towards rationalism over realism, and are more concerned with developing a perspective that acknowledges a global community than either realism or rationalism seem to be.

The authors teach international relations and Australian foreign policy in the School of Australian and International Studies at Deakin University. Gary Smith wrote Chapters 1 through 4, 6, and 7; Dave Cox wrote Chapters 5, 9, and 10; and Scott Burchill wrote Chapter 8.

Part I

International Politics and Australian Foreign Policy

1   The Transformation of the International System

INTRODUCTION

With the end of the 1980s came the end of the cold war — a dramatic, unexpected, and unpredicted closure of the main fault-line of international conflict since the end of the 1940s. A succession of communist governments in Eastern Europe were overthrown by their citizens who felt that, in Mikhail Gorbachev, the USSR had produced a leader who no longer had the will to enforce its hegemony over its post-Second World War sphere of influence. The wall around West Berlin, the most dramatic symbol of the cold war, was pulled down in 1989. The states on the Baltic Sea — Latvia, Lithuania, and Estonia — which the USSR had once incorporated into its territory, declared and achieved their independence. After an unsuccessful coup attempt against Gorbachev, Russian leaders renounced communism, the ideology under which the USSR had operated since its creation in the revolution of 1917. The USSR then disintegrated into twelve of its constituent republics, each becoming an independent state. The massive Russian Federation remained at the centre, but in the east and the south, many new states were formed: Ukraine, Belorussia, Moldova, Georgia, Armenia, Azerbaijan, Uzbekistan, Kazakhstan, Kyrgyzstan, Tajikistan, and Turkmenistan.

The 'East–West' conflict was over; maps of the world were redrawn to accommodate the break-up of the USSR. But the new conceptual maps were not easy to delineate. The 1990s appeared to be a decade of 'uncertainty'. There were few familiar signposts to the kind of international world we — as individuals or as governments — will face at the start of the twenty-first century (Leaver & Richardson 1993).

THE COLD WAR AS HISTORY

The cold war had begun by 1947, two years after the end of the Second World War. The states that had fought together to defeat the aggression of Germany, Italy, and Japan divided into two opposing groups, or blocs, once their enemies were defeated. Wartime allies became adversaries. On one side were the communist states: the USSR and the countries it controlled in Eastern Europe, including the eastern part of Germany. In China a revolution brought the communists to power in 1949. The new leaders of the People's Republic of China (PRC) then formed an alliance with the USSR. On the other side was the USA, which formed an anti-communist alliance with the states of Western Europe, including Britain, France, and the western part of Germany. Australia and New Zealand also became United States allies, signing the ANZUS treaty in 1951 and joining the South-East Asia Treaty Organisation (SEATO) in 1954.

The conflict between these blocs was called 'the cold war', a title that remained ambiguous, as it could be interpreted in either of two ways. 'Cold' could mean that relationships were 'not warm', 'chilly', 'on ice', and that optimists might look for a thaw in relationships. But 'cold' could also mean 'not hot'. A fundamental feature of this forty-year standoff was that the two sides actively sought to avoid direct military conflict with each other's military forces, as a 'hot' war might escalate into mutual destruction as a result of the development of nuclear weapons.

The cold war was a protracted conflict with several key features:

- competition for influence around the globe, and military interventions in states as part of this competition
- a continuing arms race based on the development of new and more powerful weapons of mass destruction and intercontinental ballistic missiles (ICBMs)
- extensive mobilisation of resources to sustain this arms race, and
- a systemic conflict between capitalism/liberal-democracy and Marxism–Leninism, represented by these ideologies' respective leading protagonists.

There were periods of acute tension, but fortunately the weapons of mass destruction were not launched.

The USA and the USSR were the world's two superpowers, and they attempted to organise the rest of the world around a *bipolar* structure of power, in which other states would join or associate with one of their alliance systems. The cold war began with the division of Europe, when the USSR, after driving back the invasion of Nazi Germany, retained control of the states in Eastern Europe that it had liberated. The USA and Western European states, fearing further westward expansion in Europe, formed the North Atlantic Treaty Organisation (NATO) alliance to 'contain' the USSR. The USA used economic aid (the Marshall Plan) to reconstruct Western Europe and convince the populations there that capitalism/liberal-democracy offered them greater prosperity than communism. Europe was divided into two opposing alliance systems.

The cold war extended to Asia in the early 1950s and turned 'hot' when communist North Korea attacked non-communist South Korea, and the USA and its allies, under the United Nations (UN) flag, intervened in the Korean War to protect the regime in the south. When United States forces pushed the North Korean troops close to the Chinese border, China attacked and, for the first and only time in the cold war, United States troops came into direct military conflict with Chinese troops, with significant casualties on both sides. The USA went to war again in the 1960s — this time with far fewer allies — to support the South Vietnamese government against communist forces in North and South Vietnam. The Korean War resulted in a cease-fire in 1953 and an uneasy peace; the Vietnam War resulted in the USA's withdrawal and the victory of Vietnamese communist forces in 1975. Australia joined the USA in both wars.

The nuclear arms race brought about quantum shifts in the destructive capability of both superpowers' arsenals. The USA had developed the atomic bomb in the last stages of the Second World War, and dropped two on the Japanese cities of Hiroshima and Nagasaki, prompting the unconditional surrender of Japan in 1945. The USSR tested its first atomic bomb in 1949. The USA tested the much more powerful hydrogen bomb in 1952, and conducted tests with the explosive force of up to 17 million tonnes of TNT — 1300 times the power of the bomb that destroyed Hiroshima.

In 1953 the USSR tested a hydrogen bomb, and both sides continued to test nuclear warheads until the end of the cold war. In the late 1950s the 'missile age' began, and by the end of the 1960s each side had many hundreds of intercontinental and short-range missiles equipped with nuclear warheads. Long-range nuclear-armed missiles were placed on submarines. As the 1970s progressed, multiple warheads were developed, independently targeted to leave their booster rocket in space and land on separate military installations or cities. With the missile age came the end of the traditional understanding of defence as the protection of one's territory from attack. The missiles of each side — or the sufficient proportion of them that were in hardened sites or under the sea — would survive a first strike by the other and enable massive retaliation. This system was called 'mutual assured destruction' (or MAD). The system incorporated no physical defence, only deterrence — a hope that mutual fear would prevent a nuclear war. Australia participated in nuclear strategies through the construction of United States bases: for submarine communications and as satellite ground stations for the 'eyes and ears' of the United States' weapons programs.

In the 1980s, in an attempt to achieve technological advantage and return to traditional defence, the United States President, Ronald Reagan, spent billions of dollars on the Strategic Defence Initiative project — known popularly as 'Star Wars' — to develop land- and space-based systems that would intercept and destroy incoming missiles. Critics of the project maintained that, if the project ever looked like having the prospect of success, it would be seen by the USSR as highly provocative, as the USA would then have achieved the capacity to strike first and 'win'. This, it was said, might encourage a pre-emptive strike by the USSR. It would destabilise the 'balance of terror'.

Military interventions; an arms race with massively powerful weapons; armchair strategists discussing the probabilities of global war, of controlling or not controlling a nuclear war, of a 'nuclear winter' on the planet caused by an atmosphere filled with the rubble and dust resulting from a 'nuclear exchange' — these were some of the terrifying characteristics of the cold war.

There were periods of *détente* when relations between the superpowers improved: in the mid-1950s after the death of the Soviet leader, Joseph Stalin, and most notably in the early 1970s, when some arms control agreements were concluded. United States–Soviet relations deteriorated rapidly in the late 1970s after the Soviet invasion of Afghanistan, and tensions were high for much of the 1980s, a period sometimes called the 'second cold war'.

Attempts by the superpowers to organise other states into two competing alliance systems were never fully realised, and bipolarity was challenged by the emergence of other powers, creating a tendency to a more multipolar system. Many new states, such as India and Indonesia, which had been formed out of nationalist struggles for decolonisation, declared their non-alignment in the cold war. The Non-Aligned Movement (NAM) sought in the 1950s to become a third force between the two superpowers. Some members of NAM, however, were more 'non-aligned' than others. That is, some states tilted consistently towards one of the superpowers or the other, and unity was difficult to achieve. Moreover, non-alignment as a principle was of little use in moderating conflict *between* non-aligned states, and the group of states lacked a well-defined agenda in international relations.

The most significant change in cold war power relationships, prior to the dramatic changes in the USSR under Mikhail Gorbachev's leadership, was the great split in the communist world: the rupture between China and the USSR. China shifted from being the USSR's ally in the 1950s to become a principal adversary, with a period of intense conflict verging on war in the late 1960s. Tension continued right through until the end of the cold war. From the 1960s on, the two communist powers competed for the ideological support of other communist parties around the world. The split was at least partly caused by China's desire for nuclear weapons and the reluctance of its Soviet ally to provide the technology. China tested its first nuclear bomb in 1964, after its break with the USSR. In the early 1970s, China and the USA re-established diplomatic relations. It appeared to the USSR that China had joined the USA's side in the cold war and was assisting the USA in the 'encirclement' of their country.

But there was not complete unity on the Western side. France, in the 1950s, set about developing its own nuclear weapons, its '*force de frappe*', and left its military alliance with the USA (NATO). Japan and West Germany, the defeated wartime powers, signed up as allies of the USA and maintained low-key foreign policies for several decades. They built up their economies along capitalist lines, and with less military emphasis than the USA, their economic growth proceeded at such a pace that they came to rival the economic dominance (or hegemony) of the USA. A question often asked about the post-cold war world is 'Do these challenges represent an economic axis of conflict which is replacing the old ideological and military divisions?'.

An enquiry into what the cold war was really 'all about', and how to best understand it, can now be framed as a historical question about the period 1946–89. This is a challenging task, and one subject to competing interpretations. Still more difficult, however, is the task of coming to some understanding of the 'realities' of the post-cold war world.

AFTER THE COLD WAR

Optimism

Two very different sets of consequences can be seen as flowing from the end of the cold war: an 'optimistic' scenario and a 'pessimistic' scenario. This dichotomy is necessarily artificial, but it provides a useful starting point for examining interpretations of the post-cold war world.

The optimistic view begins with the observation that the cold war had exacerbated, or even created, many regional and internal conflicts, as each superpower sought to support some states, or political group-ings within states, in competition with those supported by the other. They fought wars by proxy. With the removal of these outside rivalries, regional and internal conflicts may be lessened or even resolved. In South-East Asia, for example, a degree of peace and stability developed in Indo-China (the area encompassing Vietnam, Laos, and Cambodia) after decades of war in which the USA and the USSR (and China) were heavily involved. We can examine briefly the extent of that involvement during and after the cold war.

The 'Vietnam War' is the name usually given to the conflict lasting from the early 1960s until 1975 in which the USA intervened on a large scale with troops, equipment, and bombing campaigns in an effort to sustain a South Vietnamese government of dubious legitim-acy from attack by communist forces from the North and guerrillas in the South. Vietnamese communists received economic support and military equipment assistance from both China and the USSR, with support from the USSR predominating.

When the USA withdrew in the early 1970s and Vietnam was reunified under communist rule in 1975, war in Indo-China did not end. Vietnam became a Soviet ally after the United States defeat, and subsequently allowed the USSR to develop military bases (and to use facilities that the United States navy had built in the South). In 1978 Vietnam invaded and toppled the government of its neighbour, Cambodia, alleging border violations, and installed a government sympathetic to Vietnam. The USA then supported (with military aid) a coalition of resistance groups seeking to remove the pro-Vietnamese government of Cambodia, as well as maintaining and economic embargo against Vietnam. The USA's allies in this endeavour now included communist China (who feared 'encirclement' by the USSR) and the deposed rulers of Cambodia, the Khmer Rouge (a radical communist group who through massacres and starvation had been responsible for the deaths of hundreds of thousands of fellow Cambodians before the Vietnamese invasion). The USA persevered with this unusual alliance with communist groups to combat what was seen as the greater menace — a Soviet-backed Vietnam, which would allow the USSR to project its power into the Indian and Pacific Oceans with warm-water ports — and perhaps also to punish Vietnam for its humiliation of the USA.

With the end of the cold war the politics of Indo-China underwent a dramatic change. Russia ceased supplying economic aid to Vietnam and closed its military bases there. The USA and China ended their coalition with the Khmer Rouge in Cambodia. Vietnamese troops, having lost their Soviet backers, withdrew from Cambodia. Australia took an initiative in the UN to bring in a large temporary peacekeeping force to organise and supervise national elections in Cambodia. The UN supported the initiative and held elections successfully in 1993. The Khmer Rouge, having lost their great power backers, did not disrupt the elections. Furthermore, the wider cold war tension in South-East Asia between communist Vietnam and the non-communist states of the Association of South-East Asian Nations (ASEAN) evaporated. Vietnam, although still under Communist Party rule, sought membership of ASEAN, signed the Treaty of Amity and Cooperation with its non-communist neighbours, and began to open its economy to foreign capital investment.

Superpower arms reductions seemed to offer further cause for optimism. For the first time since the 1940s, there were real reductions, starting in the late 1980s, in the numbers of nuclear weapons held by the USA and by the (former) USSR. The USA removed nuclear weapons from its surface navy (though not from submarines). The widespread fear of global nuclear war in the 1980s eased with the end of the cold war, although the risk of accidents and miscalculations remained.

For many, a major cause of optimism was the possibility of the revitalisation of the UN, which had been established in 1945 at the end of the Second World War. Theoretically the UN Security Council had the power to use military force to counter aggression. But the five great powers of the time (the USA, Britain, France, China, and the USSR) gave themselves the right of veto of any resolution. During the cold war, the USSR and the USA both used the veto to keep the UN out of areas affecting their interests. (The major exception was the Korean War of the 1950s, in which the USA intervened under the UN flag because the USSR was boycotting the Security Council and could not veto the resolution. After this event the USSR attended all the meetings.) With the end of the cold war, there was the possibility that the world's great powers would cooperate more in using the UN for preventing, managing, and even resolving armed conflicts. United States President George Bush spoke of a 'new world order' in which the UN would play a central role.

Pessimism

Optimism may be the natural response to the end of a war, including a 'cold war', but a more pessimistic reading was also suggested by some of the events that followed. For while the cold war had certainly exacerbated certain conflicts; it had also suppressed many others, conflicts that soon found violent expression. With the disintegration of the USSR, which was in effect a multi-ethnic empire, a number of the new states reinvigorated their traditional ethnic disputes with neighbouring states or with groups within their state, or with remaining Russian influence. Examples include the dispute between Armenia and Azerbaijan over Armenian inhabitants in an enclave in Azerbaijan, civil war in Georgia, conflict in Moldova, and secessionism in Chechnya. The nuclear weapons that had been under the control of the USSR were now located in four separate states: in Ukraine, Kazakhstan, and Belorussia, as well as in Russia. Would Ukraine, Kazakhstan, and Belorussia forego these nuclear weapons? And if so, who would remove them and how? The prospect of nuclear materials and nuclear weapons becoming available on the black market did not seem implausible. The widespread fear of nuclear war that was prevalent in Europe in the last years of the cold war may have evaporated, but thousands of weapons of mass destruction remained, and some other states, such as Pakistan and North Korea, were pressing to join the nuclear 'club', of which India and Israel were already informal members.

Perhaps the most striking case for pessimism was the disintegration of Yugoslavia and the inter-ethnic massacres that ensued. The cold war had facilitated Yugoslavia's unity. Yugoslavia had the curious status of a communist but non-aligned state, with strong authoritarian leadership. With the end of the cold war, it disintegrated into warring ethnic factions of Serbs and Croats, Muslims and Christians, competing for territory with a hatred apparently undiluted by forty years of enforced peace and extensive intermarriage between the groups. How many other 'Yugoslavias' were waiting to explode, or implode?

The growing economic rivalry among the world's major economic powers was another cause for pessimism. During the cold war the USA had unquestioned military supremacy and leadership over its allies, and after the ruin of Europe in the Second World War, it also enjoyed unquestioned economic supremacy. In the mid-1940s the USA wrote the rules under which the international economy was to operate after the war. By the late 1960s the economic rise of Japan and West Germany foreshadowed challenges to the economic supremacy of the USA. After the USA, West Germany and Japan were the largest economies in the world by the 1990s. Without the cement of the cold war to bind these powers into the 'Western Alliance', would they allow their economic disputes to spiral out of control, as had happened among the world's economic powers before in the Great Depression of the 1930s?

The Gulf War

The tension between the optimistic and pessimistic readings could be seen in the development of the first major military conflict of the post-cold war world. In December 1990 the Gulf War was fought in the Middle East between Iraq and a coalition of forty states, led by the USA and endorsed by the UN Security Council (with no Russian or Chinese veto). Earlier that year Iraq, led by Saddam Hussein, had invaded its small, oil-rich neighbour, Kuwait. United States-led forces subsequently drove Iraqi troops out of Kuwait in 'Operation Desert Storm', causing a conservatively estimated 35 000 Iraqi deaths, compared with the roughly 230 deaths in the United States-led coalition that came about mostly through accidents (Keesings 1991). How is this war to be understood?

The USA expressed the view that ousting Iraq was putting the principles of a 'new world order' into practice and was an indication of the determination of states to use the UN in the way that had originally been conceived: to oppose aggression with a collective response. The war was seen to be the protection of the weak against the strong, and even of democracy against authoritarianism.

But sceptics could see other scripts being followed in a war in which media management was pursued as vigorously as military combat. Was this more a 'war for oil' — an attempt to maintain low prices and ready access to a resource over which the USA had lost self-sufficiency? The same principles did not seem to be invoked with regard to other acts of aggression, such as Turkey's annexation of North Cyprus or, closer to Australia, the Indonesian annexation of East Timor. Was UN support achieved more by United States economic pressure and threat than by the willing agreement of UN members? Why had the USA and West European powers armed and supported Saddam Hussein in the previous decade against Iran? Why were economic sanctions against Iraq not pursued as an alternative form of pressure?

Australia's response illustrates conflicting perceptions of the war. The Prime Minister, Bob Hawke, made a commitment to join the USA without consulting his Cabinet, Parliament, or other governments in the region. The rhetoric of commitment expressed enthusiasm for the 'new world order', but Australia limited its military commitment to minor naval support, in which the likelihood of casualties was low. The deeds did not match the words. They suggested a different evaluation of the stakes from that presented in speeches (Goot and Tiffen 1992; Malik 1992).

PERSPECTIVES

Australia's foreign policy operates in a world in which Australia is a small but not insignificant player, often described, or describing itself, as a 'middle power'. Foreign-policy-makers must try to establish some understanding of the realities of the world around them — to identify the constraints and the opportunities and so formulate achievable goals. But what are the realities of the international 'game'. The end of the cold war in 1989 and the subsequent disintegration of the USSR were not foreseen by analysts or policy-makers in the Australian government, whose job it is to anticipate global trends, although they were hardly alone in their failure. A far-reaching and unpredicted shift has taken place in world politics. International commentators refer to the 'uncertainty' of the 1990s, and this uncertainty is as much in the minds of commentators as it is a useful characterisation of the state of the world.

If the intellectual tools or frameworks through which the world has been understood have been found to be inadequate to the task, one response is to re-examine and reconsider the frameworks that were being used. In the academic study of international relations, an exploration of theories and concepts is often carried out at a high level of abstraction. Government policy-makers do not operate out of complex theoretical positions, but they do have particular frames of reference, and the facts of international relations do not simply 'speak for themselves'. The flux of events needs to be interpreted and understood in relation to conceptual frameworks and theoretical assumptions, even if the policy-maker does not articulate these. Consider the basic question: 'what are the most significant features of the international system?'. For students and policy-makers alike, there are a number of possible answers to this question — answers that are conceptually coherent and that compete with each other for attention:

1   *a system of states*, characterised by 'anarchy' — the absence of a central political authority, and the need for states to preserve their 'sovereignty' and military security, either alone or through alliances with other states

Realism

2   *a society of states*, in which anarchy is normatively regulated by a system of rules providing a degree of order

lib int

3   *a world economy*, characterised by globalisation of production and markets, interconnecting states through webs of trade and investment

4   *a global community*, in which global issues and human rights are the responsibility of the whole 'international community'.

Each of these answers can be seen as the core of a larger theoretical perspective on 'how the world works', a theory that attempts to identify the significant actors, events, and processes of world politics. Each perspective suggests different pressures on the Australian government and a different set of priorities for an Australian foreign policy. We can expand on each perspective, simplifying the varieties of view within each to highlight the contrasts between them.

1   Realism: a system of states

The approach that takes anarchy as the central feature of international politics is known broadly as 'realism'. It was an influential perspective for Australian governments during the forty years of the cold war, and is the dominant approach in the academic study of international relations in Australia. According to realists, the key feature of the international system is the absence of a powerful central authority: the system is constituted by sovereign states who maintain national defence organisations for their protection. The use of force in relations between states is not prohibited by any global institution or organisation. The basic rule is that of self-help; the basic objective is the survival of the state. States need to protect themselves from others through military forces and alliances. There is little or no room for ethical or moral principles in the formulation of foreign policy. Foreign policy is driven by external factors, by the anarchical nature of the international system. The maxims of realism are 'if you want peace, prepare for war' and 'peace through strength'. Peace may be achieved by a temporary balance of power as states juggle for power and position, but a universal or permanent peace is, on this view, an impossible, if not reckless, dream.

The pivotal event in the development of realist thought was the Second World War and the failure of liberal-democratic states and the League of Nations to counter the expanding power of Nazi Germany in the 1930s. In the USA in particular, realism developed as a critique of liberal internationalism, and non-power-centred assessments of international relations were dubbed 'idealism'.

For the realist, the main 'lessons' of the Second World War were the consequences of Hitler's appeasement in the 1930s and the failure of the allied nations to counter the expanding military power of Germany and Japan. These 'lessons' were then applied to the USA's relationship with the USSR after the war. When the USSR solidified its control over the parts of Eastern Europe it had liberated from the Nazis, the USA responded with a policy of 'containment'. Communism would be contained through politico-military alliances. Hans Morgenthau, an influential American intellectual exponent of realism through his book *Politics among Nations: The Struggle for Power and Peace* (1960), set out the case for a 'power politics' approach.

In Australia's case, there had been very little development of foreign policy outside the British framework before the Second World War. The shock of Britain's defeat at Singapore while large numbers of Australian troops were in the Middle East and Europe suggested that Australia had been excessively dependent on Britain and insufficiently concerned with its own military defence. One of the far-reaching consequences of the Second World War was that Britain lost its global empire with the rise of movements for decolonisation (India and Pakistan, for example, became independent in 1947). And, economically weakened by war, Britain would eventually withdraw its military forces from east of the Suez Canal. But Britain's demise was matched by the rise of the USA, who not only defeated Japan, but also went on to become the dominant global power in the postwar world.

In the concluding phases of the Second World War and in the immediate postwar years, Australia, under Labor governments, participated in complex international diplomacy in order to help build new international institutions (the UN). With the onset of the cold war and then the election of a Liberal government in 1949, realism became the dominant approach in Australia's foreign relations. For Australia, this involved attempts to form close associations with more powerful states — the search for military security through alliances with its 'great and powerful friends', Britain and the USA. As Britain contracted its military sphere of influence in the 1950s and the 1960s, Australia increasingly looked to an alliance with its remaining 'powerful friend', the USA, to 'guarantee' Australia's military security.

Australia's principal foreign-policy response to forty years of cold war was to formally align itself with the USA through treaties (ANZUS and SEATO) and to seek to deepen that alliance through various forms of military cooperation with the USA. Australia sought to protect itself from perceived international danger by close association with the most powerful state in the world. Much of the rest of its foreign policy was built around the 'cornerstone' of the United States alliance.

By contrast, many of the new states formed after the Second World War as a result of decolonisation (such as India and Indonesia) chose *non-alignment* as the 'cornerstones' of their foreign policies. These states did not wish to associate with their former colonial masters in a Western alliance against Soviet or Chinese communism.

Whether the United States alliance served Australia well is a theme explored in Chapter 3. The question here is what signposts realism provides for Australia in the post-cold war world. After the cold war, the USA continued to maintain its regional and bilateral alliance treaties, such as NATO and ANZUS, but the main communist adversaries against which, at least in the American mind, they were principally aimed, had disappeared. NATO extended eastwards, many East European states sought associate membership, and Russia joined the 'Partnership for Peace'. There was no new single axis of conflict to structure United States foreign policy and bind it to its allies, no clear sense of what the United States alliance might mean to Australia in the event of regional conflict between Australia and another United States ally or friend, such as Indonesia.

2 Rationalism: a society of states

The term 'international society of states' was used by Hedley Bull (1977) to highlight the series of rules that states in the 'anarchical' system nevertheless abide by to provide sufficient *order* to ensure their survival and prosperity. These rules begin with the recognition of, and respect for, each other's sovereignty and non-interference in each other's internal affairs; they extend to agreement over the conventions and procedures of trade, commerce, and diplomacy, and respect for the emerging body of international law. Rules create an extensive system of norms and expectations, against which the behaviour of states is judged. They create, it is argued, a common interest in a degree of order, which lessens the propensity for conflict suggested by realist accounts. The international society perspective is often known as rationalism, because of its emphasis on a state's rational perception of the conditions under which the system in which it is located can operate successfully.

In the aftermath of the Second World War, the ALP's Foreign Minister, Herbert 'Doc' Evatt, brought a rationalist dimension to Australia's foreign policy through an active role in the formation of the UN and in the provisions of the UN charter on trusteeship over dependent peoples. The UN General Assembly elected Evatt its president in 1948. But this rationalism coexisted with a realism that sought safety in alliances with Britain and the USA, rather than relying entirely on the UN system.

Rationalism has emerged at other periods of transition in Australian foreign policy, with the United States defeat in Vietnam and *détente* with China and the USSR in the early 1970s, and with the end of the cold war more recently. Gareth Evans, Australia's long-serving foreign minister from 1988 to 1996, justified Australia's policies in the areas of arms control, peacekeeping, environment, human rights, and aid partly on the basis that they were creating a setting, an international order, that contributed to Australia's long-term interests.

3 IPE: a world economy

Realism and rationalism focus on the international system as a system of states and are concerned with the power that is demonstrated in relationships between states or that is exercised through international institutions. An examination of the world economy and the processes of economic globalisation identifies different structures and relationships — a different kind of power, not located in the state. The 'world economy' consists of extensive trade in goods and services, flows of capital both on direct investment and portfolio investment (purchase of shares and bonds), and an international monetary system based on exchangeable national currencies. There are powerful non-state actors operating in the world economy. Transnational corporations and banks operate in most parts of the world with turnovers larger than the national incomes of many states, and with a concept of a global marketplace for goods, services, and labour.

If we examine Australia's trade since the Second World War, we can see fundamental shifts in the direction of exports (see Table 1.1 on the next page). Exports to Britain have shifted from overwhelming importance to minor significance. All of Europe, including Britain, now accounts for less than 12 per cent of exports. Exports to the USA have remained fairly constant at about 12 per cent of the total. Japan has been Australia's major export market since the mid-1960s, with other countries in East and South-East Asia picking up as Japan's share has stabilised in the last decade. The western Pacific region takes over 60 per cent of Australian exports. The 'great and powerful friends' that Australia has sought as military allies are not the states with whom Australia conducts most of its export trade. If realism has Australian foreign policy heading in one direction in the search for military security, international economic trends appear to take it in a different one.

So far we have considered a few elements of the international economy and Australian's connection through exports. A discussion of economic globalisation takes us a step further and considers the interaction between economics and politics, between markets and states, and seeks to identify the position of Australia in these interactions.

One general formulation of the economics–politics nexus can be summed up in the phrase 'trade leads to wider cooperation'. According to this view, economic globalisation leads to interdependence, providing a counterweight to the anarchical system of states in the military domain. Common interests and mutual economic benefit lessen the propensity for conflict in other areas. This is the central tenet of liberal (or 'neo-liberal') international political economy (IPE), a perspective maintained implicitly today by most economists who support the idea of free (or at least freer) trade. It is an outlook whose first political

Table 1.1 Australian exports to various destinations (percentage)

Destination	Percentage of total exports by year						
	1965	1970	1975	1980	1985	1990	1995
China	5.5	2.9	2.8	3.7	3.9	2.9	4.4
Hong Kong	2.1	1.5	1.1	1.4	1.5	3.8	4.0
Japan	16.6	27.7	30.5	25.2	26.0	26.3	23.1
Republic of Korea	0.2	0.3	1.5	2.2	3.9	5.3	8.5
Taiwan	0.6	0.9	1.1	1.9	3.0	3.5	4.6
North-East Asia (Total)	**25.0**	**33.3**	**36.9**	**34.4**	**38.3**	**41.8**	**44.6**
South-East Asia (ASEAN)	4.3	6.0	7.7	7.7	6.4	9.3	15.9
New Zealand	6.3	5.1	5.0	4.5	3.6	4.8	7.4
North America	12.4	15.8	12.5	11.9	8.8	11.6	8.1
Western Europe	35.0	23.7	17.2	14.1	15.7	15.5	11.2
Rest of world	17.0	16.2	20.7	27.5	29.5	17.0	12.8
Total	**100.0**	**100.0**	**100.0**	**100.0**	**100.0**	**100.0**	**100.0**

supporters included the nineteenth-century British parliamentarians Richard Cobden and John Bright, and whose theoretical position was first expounded by Adam Smith and David Ricardo.

Since the Second World War, Australian governments have generally been extremely pragmatic in addressing the issue of trading partners. Trade with Japan developed rapidly after both nations signed of a trade agreement in 1957, notwithstanding the bitter wartime memories of many Australians. Trade with China expanded in the 1960s, despite the cold war and the fact that Australia did not officially recognise the communist government in Beijing as the government of China, and despite China's support of the Vietnamese communists. Australia also developed trade and economic links with the USSR, and with the communist state of Vietnam after 1975, despite a United States trade embargo on Vietnam.

In the 1980s and 1990s Australian foreign ministers presented an image of Australia as a nation enmeshed in the dynamism of the fastest growing economic region of the world. It was implied that this growth would reinforce peace as well as prosperity in the region. Gareth Evans argued that:

> We should welcome the evidence that our region is becoming more open, less ideologically divided, particularly because in other parts of the world a form of closed regionalism is possibly emerging. We should welcome

the opportunity to play a part, as one nation among many, in a frame-work which is likely to be neither one-world, nor a bipolar world domi-nated by nuclear giants, nor a traditional form of power rivalry in which one's gain is always at another's expense, but a possible equilibrium in which power is shared, change is managed with political skill and the resulting prosperity is widely enjoyed.

(Evans 1988)

It is also possible that increased economic interdependence may, in some circumstances, lead to conflict between states — 'trade wars' were a feature of pre-Second World War international relations, and contributed to the Great Depression of the 1930s and to Japan's attack on the USA. Conflict tends to be the focus of 'mercantilist' (and Marxist) theories of political economy. The mercantilist approach sees trade as another form of nationalist power competition, and identifies trends among states towards increasing protectionism in world markets, particularly as other centres of economic strength have arisen alongside the USA. The liberal economic order estab-lished by the USA in the mid-1940s is being challenged. Declining hegemony and increasing protectionism is the basic mercantilist assessment of post-cold war prosperity (Ravenhill 1989), a view that is very sceptical of the idea that economic interaction leads to wider forms of interdependence and cooperation.

Certainly, there was a much publicised conflict between the USA and Japan in the 1980s and 1990s over the size of Japan's trade surplus (exports less imports) compared with the United States trade deficit. Will a recognition of common interests prevail over the desire to seek national advantage? Where does Australia stand when its military alliance partner (the USA) seeks concessions from its major economic partner (Japan) — concessions that may, if granted, lead to an increase in Japan's imports from the USA at the expense of Australia?

A mercantilist approach to IPE also focuses on the factors that support the creation of regional trading blocs at the expense of a global economy — where a regional group of states liberalises trade within the bloc but maintains restrictive measures on imports from outside the bloc. The European Economic Community (EEC), later renamed the European Community (EC) and then the European Union (EU), was based in part on agricultural protection through subsidies to local farmers and restrictions on food imports. This, in turn, has led to overproduction by European farmers and the subsequent selling of produce on world markets at below cost ('dumping') to get rid of the surplus. With wheat, for example, these subsidised sales have challenged the markets of Australia and the USA. The USA has responded with its own subsidies. Australia's wheat sales have been squeezed. (The USA also formed the North American Free Trade Agreement (NAFTA) with its neighbours, Canada and Mexico, in 1992.) From the mid-1980s onwards, Australia has played a very active role in attempting to influence the levels of subsidy or protection available to European agriculture and, out of self-interest as an agricultural exporter, has argued for free-trade as opposed to the hardening of economic blocs.

Leaving aside the problematic assertions that 'trade leads to cooperation' or 'trade leads to conflict', a more immediate policy question is what the government might do, if anything, to assist Australia in tapping into the prosperity that is potentially available. From the mid-1980s on, Australia undertook a major redirection of economic policy — from protection of domestic industry and regulated financial structure to a position of lower protection and more openness — in an attempt to improve competitiveness and export performance.

Labor governments under prime ministers Bob Hawke and Paul Keating presented the move to a 'more open' economy, as advocated by neo-liberal IPE, as the only way to prevent Australia's economic decline. This move involved financial deregulation and lowering tariff protection of industry. The Liberal Opposition in the 1980s and 1990s supported the general shift in economic policy and continued down the open economy path on election to government in 1996, leaving criticism of this policy to factions within the major parties and the smaller political parties such as the Australian Democrats. These groups argue that more openness in certain areas may create more vulnerability to external shocks, through the accumulation of excessive foreign debt, from speculative trading in the Australian dollar, or through increasing foreign ownership of assets.

In recognition of the increasingly blurred lines between diplomacy and international economic policy, the Department of Foreign Affairs and the Department of Overseas Trade were amalgamated in 1987 to form the Department of Foreign Affairs and Trade (DFAT).

4   A global community

A 'global community' perspective begins with the assumption that there is a greater global community of humankind, which can or should constrain the self-interested policies of states. In what ways may individuals be 'global citizens', with rights and obligations that are not determined simply by the particular territorial state they inhabit? The Nuremberg trials after the Second World War affirmed the principle that a higher duty than obedience to national laws exists and that 'war crimes' could be punished, even though they may not be crimes under the laws of the warring states. The UN has passed declarations on human rights, and in recent decades, many individuals and organisations have come to espouse ideas of human rights to which all states can or should be held to account. The Australian government has expressed support for the idea of universal human rights.

Intrastate ethnic disputes are a major factor in many post-cold war conflicts, leading to calls for intervention based on the protection of human rights. Intervention to contain ethnic conflict involves the assumption of universal rights — rights that not only take precedence over particular ethnic groups' claims of superiority or exclusivity, but also over the claims of states to non-interference in their 'domestic affairs'.

FROM PERSPECTIVES TO CONCEPTS

The key concepts that are used in discussing and formulating international relations and foreign policy — such as 'national interest', 'security', 'power' — do not have meanings that can be settled by a definition found in a dictionary. They are all part of a group of concepts that are 'essentially contested': their meaning is not settled by edict, but is developed and redeveloped in a process of intellectual and political debate. Each of the perspectives we have looked at attaches a different meaning to key concepts.

National interest

Governments will usually claim that they are adopting foreign policies that are in the national interest, rather than in the interest of a particular group in society. These claims may be actively disputed in societies where there is the opportunity for democratic debate. The former foreign minister, Gareth Evans, listed Australia's national interests, in very general terms, in the following order:

◆ geopolitical or strategic interests — the defence of Australian sovereignty and political independence
◆ economic and trade interests — trying to secure a free and liberal international trading regime
◆ being, and being seen to be, a good international citizen.

(Evans & Grant 1995, pp. 33–4)

The first point in the list is that emphasised by realists, the second by neo-liberal IPE, and the third reflects a rationalist, and possibly a global community, perspective. Does the order of Evans's list suggest a hierarchy of importance? Or does it simply reflect the order of historical development and the broadening of the concept?

Security

In realist discussion of international relations, 'security' and 'military security' are regarded as almost synonymous. But this assumption is challenged by those who argue that other forms of security are equally, or more, significant. 'Economic security' is a focus of IPE approaches, while 'environmental security' and 'human security' are a focus of global community perspectives. During his term as foreign minister, Evans increasingly adopted a wider definition of security, which reflected a movement away for a traditional realism. Discussing UN reform, Evans notes that:

> Two approaches seem particularly worthy of further exploration. The first is to develop the notion that 'security', as it appears in the Charter [of the UN], is as much about the protection of individuals as it is about the defense of the territorial integrity of states. 'Human Security', thus understood, is at least as much prejudiced by intrastate conflict as by interstate conflict.

(Evans 1994a, p. 9)

Power

The central concept of international relations is often seen to be power, and realism sees itself as explaining the workings of 'power politics' and the operation of a 'balance of power'. But its focus on the military dimensions of power may miss other crucial dimensions. Robert Nye

(1990) discusses the rationalist notion of 'soft power' — the capacity to get other states to willingly accept your agenda as theirs — in contrast to the 'hard power' of military capability, which may attempt to compel the same result, but with less success. Furthermore, from some IPE perspectives, realism and rationalism only catch one dimension of power: power as expressed in the relations of one state to another. What is missed is the dimension of 'structural' power, or the capacity to benefit from positions in international structures. Susan Strange identifies several global structures in which power is embedded, including:

- The global production structure: the sum of all those arrangements determining what is produced, by whom, by what methods and on what terms.
- The global financial structure: the sum of all those arrangements for governing the availability of credit plus all the factors determining the terms on which currencies are exchanged for one another.

(Strange 1988)

The relative power of Australia and Japan in, say, the South Pacific will be determined in part by the relationship of each state to these global structures (Solis Trejo 1994).

CONCLUSION: THE LIMITS OF REALISM

A basic theme of this book is that traditional realism, with its focus on anarchy, disorder, alliances, and military power, provides only a partial insight into the twenty-first-century world that is emerging out of the cold war. Realism is increasingly unrealistic in its image of the world. Governments as well as students of foreign policy are impelled to find other frameworks of analysis to interpret the economic, environmental, social, and political realities of emerging globalisation and the pressures that shape the foreign policies of governments (Cooper et al. 1993; Higgott & Richardson 1991).

Complex global issues are increasingly impinging on the way Australia defines its interests. The international system does not in any simple way predetermine Australia's responses — this depends in large part on how the international system is perceived by policy-makers. The perspectives and assumptions that inform a government's view of external realities will shape its perception of issues and the options, solutions, and resolutions that are pursued.

2 Making Foreign Policy

INTRODUCTION

The Australian government faces in two directions when formulating foreign policy. Facing outwards, it encounters the 'foreign', the 'world' — the variety of political, military, economic, and other pressures that compete to shape Australia's foreign-policy agenda and that limit the effectiveness with which any agenda is implemented. But, facing inwards, the government must also deal with the 'internal', the 'domestic' — the pressures from within Australia that actively seek to influence foreign-policy agendas, or that influence them indirectly.

An investigation of the domestic sources of foreign policy is also an exploration of the possibility of choice, and of alternatives to the particular policies that are adopted by a government. These alternatives can be analysed in an attempt to explain why a particular policy was adopted at a particular time. Policy choices can also be evaluated as 'right' or 'wrong', 'good' or 'bad'. For citizens in a liberal democracy, the capacity to define, debate, propose, and resist policy alternatives is the very essence of political life. This would appear to be as true of foreign policy as any other area of policy. But there are those who maintain that foreign policy is unlike other areas of policy, arguing, first, that the external constraints on choice are far greater in foreign policy than they may be in the areas of health, education, or welfare policy, and second, that there is little or no scope for normative or ethical considerations in debate on the content of foreign policy.

Australia is not a major power in the international system, however power may be defined. The scope for successful Australian initiatives is therefore limited. But is it so limited that Australia merely responds to events and decisions made elsewhere? And is the nature of its responses determined by its lack of power in the wider world?

23

Realist conceptions of international relations often take the argument about the external determinants of foreign policy a step further, suggesting that there are strong external constraints on all states in the system, even those that appear to be more powerful. From this perspective, the domestic determinants of foreign policy are overwhelmed by the international determinants; whether one examines Australia, China, or the USA, the most (and least) a state can pursue is its national interests, which can be defined in basic terms as survival. Survival is the universal interest of states — an interest imposed 'top down' by the system of states (and rejected perhaps only by revolutionary states). In an anarchic world, is Australia forced to define its 'national interests' in the same basic terms as other states: first to maintain physical survival and, second, to achieve sufficient economic success to maintain the capacity for physical survival over the years?

The idea that Australian foreign policy has little capacity to alter its definition of national interests is also implied in the accounts of Australian foreign policy that summarise the relatively fixed, 'objective' factors defining Australia's position in the world: its geographical setting, its small population, and its export-dependent economy.

Geographically, Australia is an 'island-continent' with the addition of Tasmania. It is located to the south of South-East Asia, with coastlines on the Indian and Pacific oceans to the west and east, the Timor and Arafura seas and the Torres Strait to the north, and the Southern Ocean to the south, totalling 36 800 kilometres. Australia also includes an assortment of small islands in several oceans, and it has a claim to sovereignty over parts of Antarctica. The land mass, excluding Antarctica, is 7.7 million square kilometres.

For the 40 000 years prior to European settlement, Australia's inhabitants were Aborigines, whose population in 1788 has been estimated to have been about 300 000. Australia is now, demographically, a country of about 18 million people in a world of 5000 million, or 0.4 per cent of the total. Australia's birth rates have fallen below the 'replacement rate' required to maintain longer term population levels, and population growth is largely sustained by immigration.

There are several economic indicators that can give us a basic understanding of Australia's position in the world. Its Gross Domestic Product (GDP) is 1.6 per cent of world GDP. Australia has the typical profile of a 'developed' economy in many areas, including a highly educated workforce. It is unusual in that it has a high degree of reliance on primary product exports from the farming and mining sectors and in that it has a high proportion of manufactured imports. Its high position on the world economic 'ladder' has slipped in recent decades.

The facts of geography, demography, and economy, when added to the anarchic nature of the international system, set limits in Australian foreign policy — limits most noticeable in times of war or conflict, which put a high demand on Australian human, material, and natural resources. 'Facts', however, do not 'speak for themselves'. They are interpreted by Australians in light of the values that Australians hold and the interests that they consider important. It is 'history' as much as 'geography', the 'internal' as much as the 'external', that has shaped Australians' responses to the world. The 'national interest' is not objectively predetermined. For, while there may be broad agreement that 'security' and 'prosperity' are two desirable national goals, this is no guide to the meaning of, and the means to, these goals.

The size of Australia and its coastline, for example, has often been assumed to cause *indefensibility* — with all those thousands of kilometres of coastline, how could Australia ever defend itself? The idea of indefensibility, when combined with Australia's geographical proximity to more populous Asia, and Australia's 'empty spaces', has produced frequent expressions of anxiety — anxiety about Australia's high exposure to possible military threats and about the reliability of allies in protecting Australia. An Australia highly threatened and highly desired, militarily defenceless and driven therefore into alliances with what Robert Menzies called 'great and powerful friends' — this has been a central undercurrent of Australian foreign policy from the nineteenth century to the 1970s, with continuing eddies in the 1990s. Those holding such views may believe that the precarious external setting overwhelms the possibility of domestic choice in foreign policy, and may see themselves as realists in outlook.

But how well founded are these beliefs? Many observers, in Australia and overseas, see Australia's *cultural* insecurity as explaining its attitude towards the prospect of external military threats. Perhaps the problem is one of perception, not of reality. For, apart from limited Japanese attacks on Australia in the Second World War when South-East Asia was overrun, there has been no military attack on Australia. In the Second World War, Japanese military planners considered Australia too difficult to invade because of the vast distances — they advised it would be 'a reckless adventure ... beyond Japan's ability'

(Joint Committee on Foreign Affairs and Defence 1981, p. 62). Australia has no traditional adversaries. To what extent have anxiety and threat exaggeration been products of Australian cultural isolation and insularity — a settler-colonial outlook projected into fantasies about potential military threats?

Australian defence thinking, especially since the mid-1980s, has undergone a 'revolution'. Defence strategists now firmly reject the simple notion of Australia's defencelessness, and instead have discovered the 'natural' defence assets of Australia' s size and location (Mack 1989). Much of Australia's coastline is so far from any other state that an attack would involve long and exposed supply lines. Beyond all the coastline is the 'sea–air gap' between Australia and any potential adversary, and it is in this area that any adversary is highly vulnerable to retaliatory action. This sea–air gap is far more defensible than the land borders that most states have. If any attack could successfully cross the gap, the terrain in northern Australia is formidable, and Australia's main population and industrial centres are located far away in the south-east.

Even if one accepts the logic of realism — that the anarchic international system forces military security to take priority in foreign policy — one does not find in realism an answer to the question of how that military security has been pursued or how it is best achieved by any particular state. These are issues that each individual state must determine for itself, through its political process.

After the Second World War Australian governments sought a formal alliance with the USA to supplement Australia's historical reliance on Britain, and in 1951 the government signed the ANZUS treaty. Theoretically, this was not the only option open to Australia at the time. A large number of other states chose non-alignment to advance their military security; Australia chose alignment. Australia's approach to physical security was shaped by domestic factors: negatively by fear and anxiety of possibly 'paranoid' proportions, and positively by a sense of shared values with the USA and Britain. Most Australians saw their values as fundamentally 'Western' or as liberal-democratic-capitalist, particularly in the context of conflict with communism. When the Liberal government of Robert Menzies debated the ANZUS treaty in Parliament in 1952, the Labor Opposition supported ratification. The subsequent differences between the foreign policies of the Liberal government and Labor Opposition over the subsequent twenty years were substantial, but they did not extend to the ALP's rejection of participation in the Western alliance through the ANZUS treaty.

The 'facts' of population and economic strength may have a significant role in 'determining' Australian foreign policy, but even here the picture is more complex and domestic values and interests shape the policies that governments choose in the end.

There is very little public debate in Australia about the optimum population of the country. A traditional threat-centred view expresses the dangers of underpopulation in a crowded world. But there appears to be a strong case that the aridity and ecological fragility of the Australian continent and its massive deserts preclude the possibility of a United States-sized settlement, and that Australia is more like Canada — a large land mass with tight limits on human settlement. The large civilisations of Asia did not spread to Australia in the millennia prior to European colonisation. This is presumably a consequence of the fundamental ecological constraints.

The size of the Australian population certainly limits the capacity to raise a numerically large military force based on 'manpower', even when this capacity is potentially doubled by including women in all military roles. But given the sea–air gap and lack of land borders, as well as the lack of traditional enemies, larger numbers of men and women in uniform may not be a key factor in Australia's capacity for military defence. Australia pursues a relatively 'hi-tech', low-personnel military structure, and its ability to function effectively depends more on the capacity of the wider economy to sustain it.

Australia's economy is far more significant in size than a comparison with the 'rest of the world' suggests. Compared with the economies of South-East Asia, Australia's GDP has been estimated to be about equal to that of all Indonesia, Thailand, Malaysia, Singapore, and the Philippines combined (Evans 1989a). Australia after European settlement developed as a 'trading nation' in which a significant proportion of GDP was earned through raw materials exports and which achieved a high average 'standard of living' in the nineteenth and twentieth centuries. But the path that it has taken has not simply been determined by the world economy and pressures external to Australia. Australian governments have made substantial choices, as much in response to domestic as international pressures. In the 1980s Australian governments made major decisions to move away from a high level of protection of Australian industries through tariffs, to active promotion of a more open and internationalised economy. This involved the phasing down of measures that subsidised industry in Australia, and led to the deregulation of the financial system. These choices are analysed in Chapter 8; the point here is that this shift in policy was not totally predetermined by the nature of the world economy. Many states have chosen to maintain a more protectionist approach to the dilemmas of international economic involvement.

Similar observations about the importance of domestic determinants can be made from rationalist and global community perspectives on foreign policy. Disarmament, environment, and human rights issues do not immediately become substantive issues for Australian foreign policy unless accompanied by strong domestic pressure group activity to make them so. It is this domestic pressure that will largely determine which 'global' issues governments address and which they do not. 'Think globally, act locally' is a slogan of the environmental movement, but without the local activism, it is unlikely that governments would respond as readily to global issues.

It is not useful to approach the determinants of foreign policy by asking what the national interest 'really' is, and by trying to uncover the permanent and objective factors that define those interests. It is even less useful to look to the government to answer the question for us. *The 'national interest' is a politically constructed and contested concept*, constructed by the state and its policy-makers, in response to internal and external pressures, to justify and perhaps guide its actions abroad. It is contested by groups within the state and within society, and externally too. What gets to be called the 'national interest' is a function of power relationships in a society. Governments will claim the authority of knowledge and expertise in the external world, and a mandate from voters. Critics will argue that the government is misguided, mistaken (or whatever), and will seek to change the government's definition of the national interest, to change the government, or even to campaign with no immediate effect, but in the hope of change at a later time.

Once the national interest is seen in this way, it is possible to conceive of ethical concerns shaping foreign policy (Keal 1992). The realist conception often leads to the formulation that Australia is a sovereign state in a system of sovereign states, where citizens pursue the 'good society' and propound a range of political and policy alternatives *within* the bounds of the state. Beyond the boundaries of the state, beyond

Australia, there is no scope or entitlement to seek to project one society's values onto another. To do so is to interfere in the domestic affairs of another state, something that Australians may not welcome if done by others. For the realist, the system of states limits the political goals that individuals can pursue in a moral as well as a practical way.

But if we acknowledge that foreign policy choices are being made all the time and that the national interest is contested, then presumably ethical factors can influence these choices. There may also be scope for defining interests in partly altruistic terms — that is, other than in terms of the protection and benefit of the Australian community. This is what Gareth Evans seems to have attempted when he listed the concept of 'good international citizenship' as an Australian national interest (Linklater 1992; Lawler 1992).

On the issue of opposing apartheid in South Africa, Australian foreign policy shifted fundamentally in the early 1970s from the position that it was a 'domestic' matter for the South Africans and not to be challenged through UN action or collective sanctions, to the position that it was a fundamental violation of human rights, which legitimated UN action and collective sanctions that attempted to force a change. To take another example, the basic impulse behind Australia's overseas aid program is humanitarian, even if more self-centred economic and strategic objectives shape and direct the details of the program.

AUSTRALIAN VALUES

One starting point in the analysis of the domestic determinants of foreign policy is to attempt to identify the character of the 'Australian people' — to ask if there are any values, beliefs, or ideologies that a large majority of Australians hold and that are reflected in their thinking and actions. Do shared values affect Australians' outlook on the world and shape the foreign policies that the Australian government adopts? For W. Hudson, 'politicians, bureaucrats and the few individuals and groups able to influence them do not compromise a caste: they come from mainstream Australia. They carry with them mainstream Australia's values and attitudes' (Hudson 1992).But what defines the 'mainstream'?

The difficulties here can be illustrated by a discussion of the far-reaching change that occurred in one set of Australian values over the twentieth century. For many Australians in the nineteenth century and the first two-thirds of the twentieth century, Australian identity revolved around a sense of the importance of being a White, British community (Palfreeman 1988). In the second half of the nineteenth century, the British settlers in Australia and their offspring developed the 'White Australia' policy, restricting further immigration to the those of the British 'race'. The restrictions were triggered by the arrival of Chinese fortune-seekers joining the rush to work on the newly discovered gold-fields, and later by the use of Pacific island 'indentured' labourers by Queensland sugar cane farmers. Racial exclusiveness (and belief in racial superiority) was a central pillar of 'Australian identity' at the time of federation and was expressed by all the major political parties. In 1901 the newly formed Australian Parliament passed the *Immigration Restriction Act* and a further Act that attempted to repatriate all Pacific islanders by 1906. Needless to say, the Aboriginal inhabitants were also excluded from the definition of 'Australian'. They did not come under the ambit of the new legislation: they were already 'here' and could not, of course, be repatriated. They were subject instead to a process of attempted cultural assimilation or obliteration.

Immigration criteria were eased after the Second World War to allow and encourage settlers from Southern and Eastern Europe to meet the labour shortage in Australian industry. But it was not until the mid-1960s that the government made fundamental moves to change the racially discriminatory basis of the policy. These moves accelerated to bring about the rapid dismantling of the policy and its replacement by a racially non-discriminatory immigration policy, one that would not automatically exclude as potential immigrants people from African, Asian, Middle Eastern, or Latin American countries, or indeed, Americans who were Black. The Whitlam government (1972–75) abolished the remaining trappings of the immigration policy that selected on racial grounds. All subsequent Australian governments and all major political parties supported the end of the 'White Australia' policy.

Since 1972, immigrants to Australia have been selected, without regard to 'race' or country of origin, on the basis of a points system, in which points are acquired principally for skills and/or reasons of family reunion. In addition there is a 'humanitarian' intake of refugees and other asylum-seekers. As a result of these changes, and of declining interest from traditional sources of immigration, the origins of

Australia's immigrants have altered in the last twenty years, with a decline in the proportion of Britons and Europeans and a rise in immigrants from countries in Asia and other parts of the world.

Changes in immigration policy have far-reaching domestic consequences, leading to changes in the Australian population structure and its age and ethnic composition. Governments have described the kind of society being created by the non-discriminatory immigration policy as 'multicultural' — an ambiguous term, which at a basic level conveys an official acknowledgment of, and respect for, the growing cultural diversity in Australia.

Two themes emerge from this brief discussion. First, values and beliefs change, and an attempt to identify values is not the search for some eternal truth about Australians. Yesterday's mainstream can become tomorrow's dry creek bed of discarded belief. Second, although the changing values of the Australian people have, to some extent, underpinned a change in immigration policy, it was also clear that Australian policy-makers were well 'ahead' of many sections of the population when they made the reforms.

Part of the reason for the political parties' willingness to take a lead on this issue was the external hostility to the policy as much as changing domestic opinion. In the first half of the twentieth century, the wider foreign-policy implications of the 'White Australia' policy were minimised by the fact that Australia had very little international status beyond being a dominion of the British Empire. The greatest international 'pariah' state of the post-Second World War period was the White minority government of South Africa with its apartheid system. Condemnation of White racism in Southern Africa was a rallying point for many newly independent states. Australia and South Africa had both once been British dominions. Australia's discriminatory immigration policy was seen by some states as odious, and like South Africa, as demonstrating White racism deserving of international condemnation. In the 1960s, pressure on Australia also came from the USA, where black Americans had successfully campaigned for their civil rights.

Australian business groups with international markets provided a link between external pressure and policy change. Such groups asked how Australia could hope to develop substantial economic relationships with countries in Asia and elsewhere if overseas political and business leaders found Australia's immigration rules offensive.

PARTIES, ELECTIONS, AND FOREIGN POLICY

Party traditions

Another pathway into an inquiry about domestic pressures and foreign policy is to explore the traditions of the major political parties as they have sought to define and articulate foreign policy on behalf of Australians. For most of Australia's political life since federation, Australian electoral politics has revolved around two major organised groups: the Australian Labor Party (ALP) and an anti-Labor coalition, currently composed of the Liberal Party and the National Party (formerly the Country Party). To what extent do the political parties embody different traditions in the formulation of foreign policy? To what extent do the major parties give expression to different clusters of values and beliefs in Australian society?

Political parties have an organisational structure based on State and local branches, and select candidates to stand for election to Parliament. The parties have processes whereby the organisation, on the one hand, and the successfully elected parliamentarians, on the other, collaborate (and compete) to generate 'party policy' on a range of issues, including foreign policy. The ALP has given a more powerful role to the party organisation, whereas the Liberal Party has given a greater role to parliamentarians. When in government, political parties supply, from their elected parliamentarians, a prime minister, a foreign minister, and a Cabinet, who are charged with making, or authorising, major foreign-policy decisions. In Opposition, a political party has the forum of Parliament to challenge the government and articulate alternatives.

Since the Second World War, there have been three periods of Labor government and three periods of Liberal-led coalition government (See Table 2.1). The view that there is a Labor tradition in foreign policy is based largely on the achievements of the extraordinary figure Herbert 'Doc' Evatt, Australia's foreign minister from 1942 to 1949, a period encompassing the most difficult years of the Second World War through victory and postwar reconstruction to the onset of the cold war. Evatt, a brilliant lawyer, had been appointed a High Court judge at the age of thirty-six. He resigned from the Court after ten years to stand for Parliament in 1940. He was both Attorney-General and foreign minister until the ALP lost the 1949 election. He later became a controversial and unsuccessful leader of the ALP in Opposition. But it was his period as foreign

Table 2.1 Australian prime ministers and foreign ministers, 1949–1996

Period	Goverment	Prime Minister	Foreign Minister	
1942–49	Labor	John Curtin Ben Chifley	Herbert Evatt	*7*
1949–72	Liberal	Robert Menzies	Percy Spender Richard Casey	
			Garfield Barwick	*23*
		Harold Holt	Paul Hasluck	
		John Gorton Bill McMahon	Nigel Bowen	
1972–75	Labor	Gough Whitlam	Gough Whitlam Donald Willessee	*3*
1975–83	Liberal	Malcolm Fraser	Andrew Peacock	*·8*
			Tony Street	
1983–96	Labor	Bob Hawke	Bill Hayden	
			Gareth Evans	*13*
		Paul Keating		
1996–	Liberal	John Howard	Alexander Downer	*11*

Years in power post WW2   *ALP 23*
Lib 41

minister that is seen as setting the foundation of a Labor foreign-policy tradition.

Evatt played an energetic role at the San Francisco Conference in 1945 which led to the formation of the UN. He proposed a series of amendments to the UN Charter — to limit the right of veto of the 'great powers' in the debates and resolutions of the Security Council, to expand the powers of the General Assembly, to set out the obligations of states who administered trusteeships, and to expand the scope of the Economic and Social Council of the UN to incorporate full employment as a UN objective.

In recognition of his role in the formation of the UN, Evatt was elected President of the General Assembly for 1948–49, an office that he held while continuing as Australia's foreign minister. The ALP governments were much more inclined than the Liberal Opposition to see the postwar demands for decolonisation as a positive force. The issue came close to home with the struggle of Indonesian nationalists for independence. The Dutch, who had been an Australian wartime ally, returned to assert control of its colonies in the 'East Indies' after the defeat of Japan. Conflict between the Dutch and the Indonesian nationalists intensified in the 1940s, and Australia came to play a significant role in facilitating the independence of Indonesia in 1949.

After its election loss in that year, the ALP did not return to office for twenty-three years. When it did so, a new prime minister, Gough Whitlam (who was initially also foreign minister), undertook a series of foreign-policy initiatives that created a sense of dramatic change: withdrawal of remaining troops from Vietnam, abolition of military conscription, diplomatic recognition of China, establishment of diplomatic relations with North Vietnam, opposition to apartheid in the UN, and complying with UN sanctions against Rhodesia. The brevity of the Whitlam period accentuated the sense of foreign-policy change.

The third period of post-Second World War Labor government, from 1983–96, did not begin with the same sense of dramatic change as the Whitlam years, but it is noted for a series of diplomatic initiatives undertaken by Bill Hayden and especially by Gareth Evans from 1988. Spanning these three Labor periods of government, the Labor tradition, according to Greg Pemberton, has the following elements:

(a) independence more than dependence in the conceptualisation and practice of foreign policy;
(b) regional (ie Asian-Pacific) rather than a (British) imperial focus;
(c) an internationalist more than an alliance-oriented approach to international disputes;
(d) a peaceful, conflict-solving rather than a military or power-oriented approach to international conflict;
(e) support for trusteeship and/or self-determination rather than for colonialism or the imperial or neo-colonial paths to decolonisation;
(f) and inclusive or more democratic approach to the Australian community's role in foreign policy.

(Pemberton 1994)

However, the three periods of Labor government also coincided with periods of great change in the international system: the end of the Second World War, the major cold war *détente* with China and the USSR in the early 1970s, and the disintegration of the USSR in the late 1980s. These dramatic changes in the structure of international power involved extensive consideration of the 'new world order' that might be emerging, and of 'uncertainty', 'transition', and 'optimism'.

The first two periods of Liberal government coincided with periods of greater cold war tension, during which there was greater external pressure put on Australia to align with the Western side. The first twenty years from 1949 were the classic cold war years, with the Korean War in the early 1950s and the Vietnam War from the mid-1960s. Strong intimations of a period of *détente* emerged only in the early 1970s as the USA withdrew from Vietnam and sought a *rapprochement* with China and arms control agreements with the USSR. The Fraser government (1975–83) saw a resurgence of cold war tension between the USA and the USSR, especially following the Soviet invasion of Afghanistan in 1979 (the 'second cold war'). United States voters then installed the more 'hawkish' Ronald Reagan as president in 1980. These periods of tension strengthened the natural proclivity of the Liberal-led coalition governments to distinguish their foreign policy by their degree of commitment to the anti-communist cause and the claim of a closer relationship than the ALP with great power allies in the cold war conflict. If one follows the criteria Pemberton uses to define the Labor tradition, the tradition of the Liberal-led coalition is represented, presumably, by greater dependence, a stronger imperial focus, greater alliance orientation, and a military or power-oriented approach to international conflict.

Elections

A focus on party traditions presents a picture of continuing differences between the major parties since the 1940s. However, an examination of the role of foreign policy at elections and in election campaigns does *not* support this picture. Instead, two distinct periods can be identified: the first from 1945 to 1972, in which foreign-policy issues were significant election issues, and the second from 1975 to the present, during which elections have been fought with little or no reference to foreign policies. Since the period of the Whitlam government, it appears that Australia has had a *bipartisan* foreign policy.

Disagreement over foreign-policy issues in the first period operated at two levels: at the level of disputes over approaches to specific issues, such as whether or not to recognise the communist government of China, and at a broader level over how to conceive the nature of the cold war and to understand and respond to the communist world. The issues were complicated by the role of the Communist Party of Australia, which was active in the trade union movement — the ALP's traditional support base — and by sharp divisions within the ALP on attitudes to communism.

1949 election: change of government

In the 1949 election, which saw the ALP's defeat, the Liberal Party campaigned partly on the domestic issues of excessive postwar regula- tion, opposition to the ALP's attempt to nationalise the banks, and the promise 'to put value back into the pound'. Foreign policy became an issue with the Liberals' proposal to ban the Communist Party of Australia. The ALP's alleged links to international communism became a recurring theme of Liberal election campaigning for twenty years.

The Liberal Party also declared that it would not recognise the communist government of China, which had come to power follow- ing years of civil war with the Chinese Nationalists (Kuomintang). The remainder of the Kuomintang fled to the island of Taiwan and for the next twenty-three years the Liberals would follow the USA in recognising this rump as the government of China, even though it controlled none of mainland China. The Labor government had not immediately recognised communist China in October 1949, as the Prime Minister, Chifley, was uncertain of the electoral repercussions.

1954 and 1955 elections: the Labor split

The ALP came close to winning the 1954 elections, but its vote was possibly affected by the defection of Soviet embassy official Vladimir Petrov six weeks before the election, and by Menzies calling a Royal Commission to investigate espionage in Australia. Many Labor leaders were convinced that the timing of the defection was orchestrated to affect the ALP's electoral prospects. These events precipitated the great split in the ALP as Evatt moved to denounce the organised anti-com- munist 'Grouper' faction of the party.

The ALP split of 1955 was caused by divisions within the party on the issue of its approach to communism, and the expelled group established the Democratic Labor Party (DLP). The DLP played an effective role in keeping the ALP out of office until 1972 by instructing its supporters to give their second preferences to non-Labor parties.

After the split, foreign-policy differences between the major parties became more pronounced. The 1955 Labor Conference opposed the sending of troops to Malaya to assist the British suppress the communist insurrection. It supported the diplomatic recognition of communist China (as had Britain), and recognition of China became a major policy difference between the two major parties throughout successive elections. But the split took its toll, and the ALP suffered a substantial defeat at an early election called by Menzies for December 1955.

1966 election: the Vietnam War

Party differences on the cold war and anti-communism returned to centre stage in the 1966 election. It was an election dominated by a foreign-policy issue: Menzies's announcement in 1965 that Australia would be sending an infantry battalion to South Vietnam to combat communist insurgents. In addition, Menzies had passed conscription legislation allowing young men to be conscripted for unlimited overseas military duties. In this election, fought over the issues of Vietnam and conscription, voters were given a clear choice, as the ALP declared it would withdraw all conscripts from Vietnam and all other troops after consultation with the USA. The Liberal-led coalition won convincingly.

1972 election: the Whitlam government

The war in Vietnam did not go according to plan. The 'Tet offensive' of 1968, which brought fighting to the streets of Saigon, was a turning-point in the war, undermining United States resolve to commit more than the half million troops with which it had already been unable to win. The enthusiasms of the 1966 election had evaporated by 1969, and massive anti-war protest movements were developing in the USA, Australia and throughout the Western world. For the first time since 1949, the Liberal-led coalition began to lose the initiative on the use of foreign policy as an electoral issue. The coalition's margin slipped, and the ALP, under its new leader, Gough Whitlam, returned to within striking distance of electoral victory. In 1971 and 1972 the USA abandoned its traditional approach to China and sought *rapprochement*, without consulting the Australian government, now led by Billy McMahon. When Labor leader Gough Whitlam took an ALP delegation to China in 1971, McMahon accused him of associating with the 'enemy' and warned that Whitlam would 'would isolate Australia from our friends and others, not only in South-East Asia but in other parts of the Western World as well' (*Australian Financial Review*, 13 July 1971). Three days later it was revealed that United States presidential adviser, Henry Kissinger, had been in China at the same time as Whitlam. United States actions paralysed the coalition, and foreign policy worked against it in the 1972 election that brought the ALP to office after twenty-three years in Opposition.

Whitlam and his deputy prime minister added to the sense of dramatic change by acting as a two-person interim ministry as soon as the ALP was elected, in order to implement a series of immediate decisions without waiting for the formal appointment of the full Cabinet. Most of the proposals were expressions of long-standing Labor policy positions that had been outlined in previous years. They appeared to reflect a Labor view of the world that was strongly at odds with the previous Liberal governments. On the other hand, as Whitlam himself often stated, the international environment had shifted dramatically in favour of the kind of proposals the ALP was making and implementing. Most significantly, the United States *rapprochement* with China and *détente* with the USSR produced a respite in the cold war, which the ALP's initiatives appeared to be both anticipating and responding to, and which would have forced changes in Liberal policy had they been re-elected in 1972.

Elections since 1972: bipartisanship?

In elections since 1972, bipartisanship seems to have replaced foreign-policy differences. One measure of bipartisanship is substantial continuity of foreign policy when the government is changed at an election. The three shifts of governing party after the 1972 Whitlam victory were in December 1975, March 1983, and March 1996. At each shift, the new government maintained what was, to some, a surprisingly high degree of foreign-policy continuity with its predecessor.

The Fraser government, elected in 1975, accepted completely the Whitlam government's recognition of China and set out to cultivate diplomatic relations with even more enthusiasm than the ALP. It also abandoned its traditional stance on racial issues, and accepted and even strengthened the Whitlam government's approach, particularly in opposition to apartheid in South Africa and to the seizure of power by the White minority in the British African colony of Rhodesia.

However, there was still one area of traditional difference. Fraser took a 'hawkish' line on the USSR (but not China), and at a time when the USA was still committed to *détente*. However, United States–Soviet relations deteriorated in the late 1970s, and the Soviet invasion of Afghanistan led to the 'second cold war'. There were some differences between the parties at the 1980 election on whether to boycott the Olympic games being held in Moscow in that year.

But if the Liberals developed the ALP's foreign-policy initiatives on China and racial issues, Bob Hawke, as the leader of the incoming Labor government in 1983, dressed up in the Liberal's suit of professed closeness to great and powerful friends. When the ALP took office, Hawke emphasised the Labor government's pro-American credentials by visiting President Reagan and, almost echoing Harold Holt's 'all the way with LBJ', declaring that Australia and the USA would be 'together forever'.

In his 1983 policy speech to the Washington Press Club, Bob Hawke declared extensive bipartisanship to be a desirable feature of foreign policy.

> The essential elements of Australia's foreign and defence policy have taken on a quality of bipartisanship inconceivable before [1972] … The great questions of Australia's relationship with the United States, the People's Republic of China, the Soviet Union, the European Economic Community, Indonesia, our special relationship with the Commonwealth of Nations, Papua New Guinea, New Zealand and Japan, and our conduct on Southern African questions now possess a continuity, consistency and consensus.
>
> (*Commonwealth Record*, 1983, p. 216)

The Howard Liberal government took office in 1996, amid considerable scepticism in parts of Asia about whether the new government would give the same high priority to the 'region' as the ALP had done. Some of the coalition's statements while in Opposition suggested that the ALP had given an excessive priority to the region over Europe and the USA. In the election campaign, Paul Keating presented the ALP as the only party with the commitment to securing Australia's interest in the region. Once in office, however, Howard and his foreign minister, Alexander Downer, set out on active regional diplomacy, with the avowed aim of maintaining the ALP's commitment and momentum.

INTERPRETING BIPARTISANSHIP

One explanation of the post-1972 bipartisanship is that the 'experience of office' creates a pressure for continuity — that governments with a variety of foreign-policy ideas formed in Opposition quickly discover a new perspective, and new constraints, when forced to look at issues from the point of view of Australia's government. As a government they are responsible for the consequences of policy and they find themselves accepting the policies of their predecessors. But an examination of elections since 1972 also suggests bipartisanship at a deeper level: a convergence between the policies of the government of the day and the Opposition of the day. There have been almost no foreign-policy issues *raised* by either party at any election since 1972.

Many commentators see the Whitlam government as standing at a great historical juncture between two periods. Prior to Whitlam was the epoch of conflict over foreign policy; after Whitlam came the great epoch of bipartisanship extending to the present day. This transition is seen as analogous to individual development — as a growing 'maturity' (Mediansky 1992), a move from 'dependence to independence' (Bell 1988), a 'diplomatic coming of age' (Bull 1987). These depictions suggest a complacency with the status quo — a simple line of progress from the past, with incremental self-evident changes to come, no doubt with the 'wisdom' of age. Similar labels of maturity were used in the Evatt years of the 1940s, after the First World War, and at the time of federation in 1901. How many times can Australia be said to have 'grown up' in a century?

Minor parties and independents have often chafed at the restrictiveness of this bipartisanship in the face of what they see as pressing issues of human rights, environment, war, conflict, and poverty, and an excessive dependence on the USA. Bipartisanship between the major parties has been most vividly demonstrated when the Australian Democrats and independents have initiated legislation and resolutions in the Senate, and the Labor and Liberal parties have combined to defeat the Bills.

To what extent is there still a substantial degree of debate and disagreement *within* the major parties and among their supporters that does not find expression at the apex of the system? Trevor Matthews and John Ravenhill (1988) conducted a detailed analysis of the extent of bipartisanship among a section of Australia's political and business 'elite'. They found:

- strong bipartisanship on relatively few issues, such as the political and economic importance to Australia of the USA, China, Japan, and ASEAN
- weak and nominal bipartisanship on a series of economic, strategic, political, and humanitarian issues
- the absence of bipartisanship on many strategic issues and some economic issues.

In other words, fundamental difference in outlook remained between members and supporters of the two major political groupings. For example, the Labor MPs surveyed were much less inclined to support the propositions that Japan should develop a larger defence force. They were also much more sceptical of the proposition that ANZUS would lead to the USA's assistance of Australia in a crisis or could help to solve international problems, or that multinational corporations would help solve regional and international problems. ALP supporters interviewed believed that export restrictions on conventional arms and nuclear disarmament were in Australia's interests — propositions rejected by a majority of coalition supporters surveyed at the time.

FOREIGN-POLICY DECISION-MAKING

All political parties, and any individual or group, can enunciate foreign-policy positions and preferences. But foreign policy 'proper' — that is, the foreign policy of the state — is elaborated by the party or parties in government. The government, in turn, is subject to the procedures and rules of Parliament, where its right to exist as a government is established by the support of a majority of votes in the House of Representatives. A large array of influences and pressures can come to bear on a foreign-policy decision, and the nature of these pressures also varies with the kind of foreign-policy decision being made. Foreign-policy decisions can include the following:

- multilateral diplomatic initiatives, which join a number of other states in some common purpose
- bilateral diplomatic strategies, which manage short- and long-term relationships with another state in order to pursue particular objectives
- military initiatives, such as going to war, joining a military alliance, or joining a multinational peace-enforcement or peace-keeping force
- symbolic actions, such as endorsements of resolutions, statements of support, or denunciations, which require no other action (all decisions have a symbolic dimension, and symbolism is particularly important in diplomacy, where offence can easily be taken and misunderstanding is an occupational hazard)
- national rules for interacting with the international system, such as foreign-investment guidelines, tariff levels on imports, export controls, or immigration criteria.

Parliament

In Australia, the Parliament has only a small role in foreign policy compared with what is traditionally called the Executive — government ministers and their departments. Most foreign-policy decisions, even decisions to go to war, do not require an Act of Parliament. The Senate, not controlled by the government, has taken the lead in proposing parliamentary inquiries into areas of foreign policy, either through its committees, such as the Senate Standing Committee on Foreign Affairs, Defence and Trade, or in conjunction with the House of Representatives. These committees hear submissions from, and direct questions to, private individuals and 'experts', members of non-government organisations, and departmental officers. Transcripts of evidence are available, and one or more reports are usually produced and published after each inquiry. Some recent parliamentary inquiries have been held on:

- Australia and Latin America
- Australian–Indian relations
- Australia's role in UN peacekeeping
- Japan's defence policy
- Australia and the South Pacific
- Australia and Southern Africa.

Parliament could play a decisive role if a foreign-policy issue was to lead to defections or resignations from the government, which in turn could remove the ruling party's majority in the House of Representatives and topple the government. This happened in 1942, during the Second World War, when the Menzies government was removed from office by the shift of allegiance of two of its previous supporters in the House of Representatives.

The informal and relatively powerless role of the Parliament in day-to-day foreign-policy matters is in sharp contrast to the United States system, in which power is divided between Congress (the parliament) and a separately elected president. Powerful congressional committees, such as the Senate Foreign Relations Committee, play a major role in foreign-policy formulation and the selection of key officials. Australian parliamentary committees can only hope to contribute indirectly to decision-making by raising the level of public information and debate on particular issues.

Prime minister and foreign minister

Among Australian government ministers, the two most important with regard to foreign policy are the prime minister and the foreign minister. Scott Burchill writes:

> The most important individual in the decision-making process is the Prime Minister ... With ministerial seniority to the Minister for Foreign Affairs and Trade, the Prime Minister can appropriate foreign policy issues at will. From approximately 1985 on, Prime Minister Hawke took much more interest in Australia's bilateral relationships with the US, the USSR and the UK, effectively squeezing foreign minister Hayden out of these areas of his portfolio ... Hawke's decision to reverse government policy on Antarctica in 1989, from supporting the existing minerals convention to the proposal for a wilderness park, was a humiliating change for foreign minister Evans, but one which again demonstrated the Prime Minister's seniority in this area of government policy.
>
> (Burchill 1994, pp. 14–15)

Even Evatt in the 1940s found himself overruled or led by the Prime Minister, Ben Chifley, from time to time: Chifley pressed his foreign minister to give greater support to Indonesia as it sought independence from its Dutch colonial masters, and Chifley firmly aligned Australia with the USA in the developing cold war, a move that Evatt attempted to resist.

The prime minister and foreign minister are part of the Cabinet, and Cabinet has been described by Hugh Smith as at 'the heart of the Executive' (Smith 1992, p. 23). The Hawke and Keating Cabinets, however, considered foreign-policy decisions somewhat rarely. The decision to support the use of force in the Gulf War in 1990 was taken by the Prime Minister in discussion with a few senior Cabinet colleagues, but without referral to the Cabinet.

The ALP has a biennial conference, which outlines the broad policy 'platform' of the party. The conference is composed of federal and state parliamentary leaders, and elected delegates from the state and territory branches of the party. After the election of the first Hawke Labor government in 1983, Hawke made several foreign-policy decisions that were in conflict with conference resolutions and the party platform, on uranium mining and on the Indonesian incorporation of East Timor. The 1982 platform, formulated when the ALP was in Opposition, committed the party to a moratorium on uranium mining and repudiation of commitments entered into by previous governments. It also proposed a plebiscite by the East Timorese to determine whether they wished to remain under Indonesian sovereignty.

At a fiery conference meeting in 1984, the party's platform was altered to reflect the new Prime Minister's decisions. Uranium mining would be allowed to continue, and the resolution on East Timor was diluted. This conference was accompanied by dramatic resignations from the party by disappointed members and the burning of ALP membership papers. The Nuclear Disarmament Party (NDP) was created, winning Senate seats in the 1984 and 1987 elections. In 1985 Hawke announced that the Australian government recognised the sovereignty of Indonesia over East Timor.

The meeting of all federal parliamentarians within the party (the 'Caucus') theoretically has the power to review all Cabinet decisions, but it is now relatively rare for foreign-policy issues to be discussed here. The ALP operated in the Hawke–Keating period with two well-organised factions — the 'Right' and the 'Left' — and with one less organised faction, the 'Centre Left'. As the Right has dominated both the Cabinet and the Caucus, the factional system has tightly constrained the amount of debate in Caucus over policy issues.

Bureaucracy

One of the most significant differences between the formulation of actual foreign policy in government and the 'virtual reality' of foreign-policy proposals put forward by an Opposition is the presence of a large permanent bureaucracy that advises the ministers on policy alternatives and implements government policy. The bureaucracy is organised into departments, each under the direction of a minister and often an assistant minister. The prime minister has one department (the Department of the Prime Minister and Cabinet); the foreign minister has another (the Department of Foreign Affairs and Trade, or DFAT). But who really 'has' whom. The bureaucracy is often seen collectively as the major influence on foreign-policy-making, due to its access to information, expertise, and historical experience, and to the time and resources it can put into formulating foreign-policy positions. Ministers come and go; departmental secretaries and deputy secretaries have longer life spans. Besides PM&C and DFAT, other key departments involved in the formulation of foreign policy would include Defence, Immigration, Treasury, Finance, and the Department of the Attorney-General.

Ministers also have personal advisers in addition to the public servants in their departments. The policy-making arena this creates has been heavily satirised in the television series 'Yes Minister' and 'Yes Prime Minister'. 'Bureaucratic politics' — the extent to which departments have formulated a 'line' on a particular issue and brokered its acceptance with competing government departments — may be crucial in many foreign-policy areas. One such area is Australia's relations with Indonesia and, in particular, Australia's response to the Indonesian invasion of East Timor. This response is to take the 'line' first expressed by the former ambassador to Indonesia, Richard Woolcott, which maintains that 'good relations' with the Indonesian government is so important that Australia should recognise the incorporation of East Timor, de facto and de jure. Liberal and Labor governments have accepted this line. Since the massacres in the East Timorese capital of Dili in 1991, Australian foreign ministers, while not shifting from this basic position, have come to emphasise the need for greater autonomy in East Timor. By way of contrast, the departmental line in DFAT in favour of signing an Antarctic Minerals Convention was dramatically overturned by the then Prime Minister, Bob Hawke, without informing either the department or the Minister, in order to shore up the environmental vote at the 1990 election (Elliot 1993).

Pressure groups

Into this world of party traditions, ministers, party organisations, bureaucracies, and personal advisers enter interest or pressure groups seeking to change, alter, or reinforce foreign policy on a range of matters, from the broad and general to the highly specific. Key questions, on issues such as how much power pressure groups exert and how appropriate the exercise of this power is in a liberal-democracy, cannot be easily answered.

Paul Dibb, introducing a collection of essays on Australian foreign policy, claimed that 'Politicians too often approach Australia's external relations in an opportunistic manner, responding to the day-to-day demands of domestic pressure groups or influential lobbies in the private sector, trade unions or other sectional interests (Dibb 1983, pp. 12–13). On the other hand, Anthony Bergin concluded his study on the subject of pressure groups and foreign policy with the claim that, 'In surveying pressure groups and their influence on the foreign-policy process, it may be concluded that their general influence seems slight (Bergin 1983, p. 14). A pressure group or interest group can be defined as any group seeking to influence any aspect of public policy, including foreign policy. This vast array of groups can be classified in a number of ways: as sectional or promotional, open or closed membership, insider or outsider groups, and by the techniques they use in attempting to influence policies.

One distinction is between groups that are sectional or self-interested and those that are 'promotional' or altruistic. Sectional groups often have closed membership, such as the Australian Mining Industry Council (AMIC) or the Australian Council of Trade Unions (ACTU). Ethnic organisations also fall into this category. Altruistic groups often promote some idea of the 'common good', whether it be for 'Australia', for humanity at large, or for some group that lacks a political voice, such as victims of torture.

Pressure groups can be classified by the type of issue on which they focus and on the grounds of whether they have a long-standing interest in an area of foreign policy or simply a concern for an issue of the moment. Table 2.2 is illustrative only and makes no attempt to be comprehensive or to suggest that any particular group is successful in achieving its aims.

There seems to have been considerable expansion in pressure group activity in the last decade, confirming Dibb's impression. But this does not necessarily confirm his judgement that politicians 'too often'

Table 2.2 Examples of Australian pressure groups

Type of organisation	Organisation	General issues	Particular issues
economic	Australian Mining Industry Council	investment rules	uranium policy
	Australian Chamber of Manufacturers	tariffs	export assistance
	Australian Council of Trade Unions	worker rights	
ethnic	ethnic community councils	racial vilification laws	recognition of new states
environmental	Australian Conservation Foundation	climate change	Antarctica
	Wilderness Society	forests policy	woodchip exports
human rights	Amnesty International	torture	individual victims
	East Timor independence groups	self-determination	East Timor
Third World development	Aust. Council for Overseas Aid	poverty	famine
peace	United Nations Association	arms control	Bougainville war
	Medical Association for Prevention of War	conflict resolution	

respond to pressure groups. In a liberal-democracy, pressure groups can be seen as playing a central role in articulating wider community concerns, developing public debate on issues, and pressing for change in policy (or to maintain the status quo). Of course some groups may seek to stifle public debate or manipulate opinion. This brings us to a central issue in liberal-democratic political theory. According to some accounts, it is the very pluralism of group activity that prevents the emergence of authoritarian rule. This implies that the capacity to organise is evident throughout society and that 'checks and balances' operate in the struggle for influence over government policy. A pressure group can be seen as exerting 'undue' influence if its role is out of all proportion to its membership or community support (for example, if its influence is based principally on financial clout). Hugh Smith (1992, p. 34) observed that, in Australia, there 'is no military-industrial complex or a Jewish lobby on the scale that exists in the United States'.

In some areas the government has created formal advisory groups to coordinate advice received from pressure groups. In recent years these have included the Trade Development Council, the Trade Negotiations Advisory Group, the National Consultative Committee on Peace and Disarmament, the Australia–Japan Foundation, and the Australia–Indonesia Institute (Evans and Grant 1995, p. 50).

Australian governments acknowledge Australia to be a 'multicultural' society, and this acknowledgment points to the possibility of ethnic mobilisation on foreign-policy matters. The government's 1994 decision to recognise the new state of Macedonia, but not to use Macedonia as the name of the state, led to a dramatic reaction by immigrants from the area, some of whom saw it as evidence of the influence of the 'Greek lobby' on the government. With the end of the cold war, and with the disintegration of the multi-ethnic states of Yugoslavia and the USSR in particular, ethnic mobilisation has played a more significant role in international conflict. Warnings about the dangers of ethnic pressure groups influencing Australian foreign policy have been made by several commentators. Former ambassador Ralph Harry (1982, p. 73) concluded a study on the subject with the injunction that 'We must seek to prevent the continuation of non-Australian quarrels in Australia or their injection into our foreign policy'. Peter Boyce (Boyce & Angel 1992, p. 5) has warned about 'the security risk created by immigrant Australians harbouring divided loyalties'.

Is there a tone of panic and unrealism in the sweeping character of these warnings? 'Ethnic minorities' as pressure groups are not, after all, a new feature of Australian politics. The 1916 conscription referenda were opposed by, among others, Archbishop Mannix, leader of a Roman Catholic congregation of largely Irish origin or descent, at a time when Britain was at war with Irish republicans. The success of these campaigns created what is usually described as an 'Australian' tradition of opposition to conscription, which was a powerful factor in subsequent wars. It is, of course, quite 'reasonable' to expect all pressure groups, ethnic and otherwise, to operate within the bounds of peaceful protest. Many immigrants to Australia have been subject to substantial persecution on ethnic lines before settling in Australia and value their new home as a safe haven. On the other hand, conflicts that have led to emigration have often involved substantial abuses of human rights, to which the Australian government might be expected to respond.

In a variety of ways, the mass media assist pressure groups in their efforts to draw attention to their case or their cause. Television images of starvation in Ethiopia affected Australia's aid from the mid-1980s, and similar images of Somalia played an important role in mobilising support in the USA, and in Australia, for UN 'humanitarian' military intervention. The mass media have also generated foreign-policy issues for the Australian government: foreign governments who expect Australian governments to be able to control the media take offence at print, radio, film, or television material produced in Australia; the Australian government has also been in occasional conflict with individual journalists and sections of the media when the government's concern for secrecy conflicts with the 'right to know' (Maher 1992; Toohey & Wilkinson 1987).

CONCLUSION

Foreign policy is not substantially different from other areas of policy in terms of the array of domestic pressures that come to bear on its formulation. Some decisions, such as participation in the Gulf War, appear to be taken by a tiny policy-making elite (although the decision on the Gulf War then had to be 'sold' to the ALP and the wider community); others involve a strong bureaucratic line, such as recognition of Indonesian sovereignty over East Timor; still others, such as the creation of an Antarctic 'world park', seem susceptible to pressure group influence. Foreign policy is different from other areas of policy in that other states have a direct interest in aspects of Australian foreign policy, and domestic pressures need to be balanced by governments against foreign reactions. Foreign policy is also different from other areas of policy in that the arena of implementation is often the world outside Australia, and the success of implementation depends on cooperation with forces outside Australia.

Part II

Identities and Alignments

3 ANZUS Alliance

INTRODUCTION

Then …

In 1951 Australia, New Zealand, and the USA signed the ANZUS treaty, formalising for Australia the 'American alliance', or simply 'the alliance'. Australia's foreign minister at the time of signature was Percy Spender, who left Parliament soon after to become Australian ambassador to the USA.

Assessing the working of the treaty after nearly twenty years, Spender wrote in his political memoirs, *Exercises in Diplomacy*, that the ANZUS treaty 'has been and remains the essential pillar of Australia's security. It is a treaty which accords while it endures the protective shield of the mightiest power in the world against any armed attack upon our country no matter from what nation that armed attack may come (Spender 1969, p. 185). This was the grand idea of the security *guarantee*, the belief in the assured protection of Australia by one of what Menzies called our 'great and powerful friends'.

If we got all, we seemed to give all. Harold Holt, in his brief tenure as prime minister before drowning at Cheviot beach, visited United States President Lyndon Baines Johnson in 1966. Australia had already joined the USA in the Vietnam War, and Holt declared to Johnson that Australia would go 'all the way with LBJ'. Five hundred Australians died in the Vietnam War.

… Now

In the Defence White Paper of 1987, Kim Beazley, the minister for defence at the time, declared that, 'While it is prudent for our planning to assume that the threshold of direct United States combat aid to

Australia could be quite high in certain circumstances, it would be unwise for an adversary to base its planning on the same assumption' (Department of Defence 1987, p. 5). The 'adversary' cannot be sure that the USA will not support Australia. We tell others that the mighty shield *may* still protect us, but in doing so, we are also telling ourselves that it may *not*. The 'guarantee' appears only to be a limited warranty of unsure utility.

More ambivalent rhetoric and more limited commitments matched these uncertain assurances. Hawke's 1983 declaration in the USA that the USA and Australia were 'together forever' appeared even more effusive than Holt in the 1960s. But it was the same Australian prime minister who, a few years later, declared that the ally's subsidised agricultural sales were a 'bullet ... in the head' and that, even if not intended, 'it hurts as much as if it were aimed' (*The Australian* 28 April 1989).

Hawke responded quickly to United States President George Bush's request for a military commitment to the Gulf War in 1990, and as with Menzies in 1965, the response was so rapid that it elicited a debate about the extent to which the request was actively solicited before it was received. Yet the substance of the Australian military commitment was small and did not match the rhetoric. While parts of the navy went to the Gulf, the infantry stayed home. There were no Australian casualties.

The Australian–United States alliance was known for many decades in Australia as the 'ANZUS alliance', until a taken-for-granted New Zealand effectively took itself out of the alliance in 1984 by banning the entry of nuclear-powered or armed vessels to its ports. The USA replied that it had suspended its obligations to New Zealand under the treaty. ANZUS came to describe two *bilateral* relations: Australia with the USA and Australia with New Zealand. And although the treaty remained, the term 'ANZUS alliance' fell into virtual disuse.

THE UNITED STATES ALLIANCE: ORIGINS AND DEVELOPMENT

The Australian–United States alliance has been constructed from two major elements:

1   the ANZUS and SEATO treaties, which set out undertakings by the 'parties' to act if one of the members is subject to armed attack or threat thereof

2   a wider pattern of military cooperation — including participation in wars, intelligence sharing, purchase of equipment, and weapons — some elements of which have also been formalised in agreements and memoranda.

The catalyst for Australia's pursuit of a United States alliance was Japan's dramatic entry into the Second World War in December 1941. This assault defeated Australia's traditional protector by destroying British power in East and South-East Asia. But the assault also brought forth hope in a new protector as it unleashed United States counter attack, which eventually drove Japan to unconditional surrender in August 1945. The Australian government collaborated extensively with the USA during the war. United States General Douglas Macarthur set up headquarters in Melbourne in early 1942 as Supreme Commander of Allied Forces in the South-West Pacific. Although Australia was not often consulted on matters of high strategy, extensive intelligence cooperation was established. This cooperation continued after the war when Australia joined Britain and Canada in the secret 'UKUSA agreement' for intelligence sharing (Richelson & Ball 1985).

After the war Australia's Labor government, preoccupied with the possibility of a resurgent Japan in the future, sought a formal alliance in the Pacific region involving the USA. Lack of United States interest meant that the matter of a treaty was not resolved until the 1950s, when the Liberal-led coalition was in office and the government was able to conclude a treaty that both the major Australian political parties had sought. The outbreak of the Korean War in June 1950 and the commitment of United States combat troops prompted the USA to extend its anti-communist containment policy from Europe to the world at large. The USA sought security treaties in other parts of the world to complement the NATO alliance with Western Europe.

The ANZUS treaty

The Australian and United States governments had different perceptions of the meaning of ANZUS when they signed the treaty in 1951. Australia's primary security concern was Japan rather than China; the USA, however, had begun to recast Japan as an ally in the containment of communism. Australia saw ANZUS primarily as reinsurance against a 'soft' peace treaty with Japan; the USA saw it primarily in the context of the cold war, as a treaty that would tie Australia and New Zealand into the global effort to contain communism.

The main provisions of the treaty are disarmingly brief:

Article II
In order more effectively to achieve the objective of this Treaty the Parties separately and jointly by means of continuous and effective self-help and mutual aid will maintain and develop their individual and collective capacity to resist armed attack.

Article III
The Parties will consult together whenever in the opinion of any of them the territorial integrity, political independence or security of any of the Parties is threatened in the Pacific.

Article IV
Each Party recognises that an armed attack in the Pacific Area on any of the parties would be dangerous to its own peace and safety and declares that it would act to meet the common danger in accordance with its constitutional processes ...

Article V
For the purpose of Article IV, an armed attack on any of the Parties is deemed to include an armed attack on the metropolitan territory of any of the Parties, or on the island territories under its jurisdiction in the Pacific, or on its armed forces, public vessels or aircraft in the Pacific.

The meaning of these Articles for Australia's security has been strongly contested in Australian foreign-policy debates over the last fifty years.

SEATO

In September 1954, Australia became party to a second pact with the USA: the South-East Asia Treaty Organisation, or SEATO. Its membership was wider than ANZUS and included Britain, New Zealand, and two South-East Asian countries: Thailand and the Philippines.

SEATO was established after the First Indo-China War, in which the Vietnamese nationalists, led by the communist-oriented Viet Minh, defeated the French. The USA interpreted events in Indo-China as further evidence of communist expansionism that was being controlled from outside the country by the USSR and China. In particular, the USA saw the treaty as a way to retain the temporary division of Vietnam that had been negotiated at a conference in Geneva in 1954, with a communist-ruled state in the north and a non-communist state in the south. Like ANZUS, SEATO stipulated that an attack upon one party would lead to action being taken by the others to meet the 'common danger' in accordance with their 'constitutional processes'. The USA added to the treaty its 'understanding' that it would act only against 'communist aggression'.

From British to United States ally

In the 1950s, Australia concluded pacts in which the principal great power was the USA. However, this did not imply a straightforward transfer of alignment from Britain to the USA. Britain was a member of SEATO along with the USA. Australia, New Zealand, and Britain were parties to the informal Australia, New Zealand and Malaya (ANZAM) arrangements from 1948, designed to coordinate the military planning of the three countries in the South-West Pacific region, including British-controlled territories in South-East Asia. The USA was not part of these arrangements. Australia sent ground troops to Malaya in 1955 to help the British suppress a communist insurgency there. It also sent troops to North Borneo (now Malaysia) in 1965 during Indonesia's 'confrontation' with the new state of Malaysia.

More generally, Australia in the 1950s presented itself in the international arena as a member of the British Commonwealth. Menzies often emphasised the value of the British connection. Australian policy tried to minimise the differences that emerged between the foreign policies of the USA and Britain. It also aimed, rather grandiosely, to bring the position of the two closer together. When differences remained, Australia usually adopted the British position — up until the Suez crisis.

One of the greatest shocks to Australian diplomacy in the 1950s was generated by the disagreement between Britain and the USA over the 'Suez crisis'. In 1956, Britain, France, and Israel secretly planned and began to carry out an invasion of Egypt in response to the Egyptian government's nationalisation of the Suez Canal Company. Strong United States disapproval led to the abandonment of the invasion. This was the most significant conflict to occur between Britain and the USA since the Second World War. At the start of the invasion, Menzies backed Britain. Initially he tried to justify the military invasion of Egypt, but the condemnation expressed by the USA and a very large majority at the UN left Australia diplomatically isolated.

For Britain, the main lesson of the Suez crisis was that it was no longer the great imperial power it had been before the Second World War. Britain lacked the economic strength to re-establish a 'great power' role. In the 1960s it began to reappraise its economic and military policies. In 1961 the British government announced its intention to apply for membership of the EEC (or the 'Common Market'). Its efforts were blocked by the French, but Britain persevered and was finally successful in its application in 1971. Over this period Britain's economic links with members of the Commonwealth, including Australia, declined rapidly. On the military side, Britain finally announced in 1967 that almost all its military forces 'east of Suez' would be withdrawn within ten years. In 1968 the timetable was shortened considerably.

In the late 1950s, and even in the early 1960s, the Australian government appeared hesitant to recognise the decline of British power. After the Suez crisis, however, Australia was less inclined to follow the British line when Britain and the USA disagreed. In 1957, Australia decided that in future it would standardise its military equipment with that of the USA, not Britain, and in 1960 Australia agreed to pursue a mutual weapons development program with the USA.

Menzies had originally characterised Australia's search for allies as a desire for powerful *friends*. By the mid-1960s, Britain had declined in power and was fast losing the will to attempt to influence the military security conditions in Australia's region. When Australia went to war alongside the USA in Vietnam, Britain did not participate. Holt's 'all the way with LBJ' was, in part, public recognition that the end of Britain's role in the region was imminent and an acknowledgment that the USA was the only great power ally that counted.

Military commitments with the USA: forward defence

Australia's participation in the Korean War in the early 1950s, and in the Vietnam War from the mid-1960s to the early 1970s, was part of the military strategy of forward defence, a policy of maintaining military forces overseas with a great power ally and, if necessary, involving them in military conflict there. This policy was thought to be preferable to planning to fight wars on Australia's territory or in its immediate approaches.

The Korean War

When North Korea invaded South Korea in June 1950, the USSR was absent from the UN Security Council. This absence enabled the USA to establish a military force under the UN flag to support the government of South Korea. However, only a relatively small number of UN members sent ground troops. Australia was one of them, and did so promptly. Such a quick response probably helped Australia in its efforts to conclude the ANZUS treaty with the USA.

Despite its promptness in sending military forces, Australia sent fewer than it could have. Consequently, the USA applied pressure on Australia to commit a second battalion of infantry. (A total of 5133 Australian were sent to the area, and 261 were killed. The USA sent 350 000 troops, and more than 33 000 were killed.) Australia's hesitancy was a result of Menzies's view that Britain might require an Australian demonstration of support, either in Malaya or, if a new world war developed, in the Middle East.

The Vietnam War

After the military defeat of the French by the Viet Minh, the Geneva Conference of 1954 'temporarily' divided the country, pending elections. The USA became the major financial prop of the new regime in South Vietnam. This regime faced a mounting insurgency in the 1950s, aided by the communist government of the North. SEATO was supposed to guarantee collective action against 'communist aggression', but it was clear by the early 1960s that other SEATO members, namely France and Pakistan, would not accept United States initiatives. As a result, the USA redefined its interpretation of the treaty to allow it to act without the agreement of other members.

In the 1960s, the USA became involved in a rapidly escalating military intervention to shore up the South Vietnamese government. In the first half of the 1960s, American troops were officially there as 'advisers' and not combat soldiers. However, from 1965 to 1967 the commitment escalated dramatically from about 30 000 to more than 500 000 troops in South Vietnam. Heavy aerial bombings of North Vietnam by the USA began in 1965 and continued into the 1970s. Despite this massive commitment of troops and firepower, victory was not achieved. United States troops withdrew in the early 1970s, and the regime in the South collapsed in 1975. There were 58 000 Americans and between two and three million Vietnamese killed in this war.

The USA involved itself in the Korean War under the UN flag. However, it could find fewer allies ready to fight alongside it in Vietnam. Of the handful of countries that supported the USA with troops, Australia's contribution was the second largest next to South Korea's. Australia's participation was justified at home partly by its obligations under SEATO. Initially only thirty instructors were sent in 1962. In April 1965, Menzies announced that Australia would send a battalion to Vietnam. The size of Australia's forces overseas was increased in stages between 1965 and 1967. As the war turned, the withdrawal of Australian troops in the early 1970s was phased in with the withdrawal of the USA.

United States facilities or bases in Australia

The stalemate reached in the Korean War and the defeat experienced in the Vietnam War reverberated in the domestic politics of the USA and Australia for decades. But as acts of military cooperation between Australia and its new great power ally, they were episodes that had a limited life-span. However, Australia and the USA also developed areas of military cooperation that were more long-standing and continue to this day: the establishment in Australia of United States 'joint facilities' (as governments have described them), or 'bases' (the term preferred by critics).

Australian governments have generally argued that agreements to establish installations are logical extensions of the ANZUS treaty. Through hosting these installations, Australia has become an integral part of United States military and nuclear strategy, both in the region and globally. The list of United States installations in Australia that are concerned with defence, science, and intelligence-gathering by technical means is considerable. Much of the information that the Australian public has on these installations, particularly on North West Cape, Pine Gap, and Nurrungar, has been gathered and published by Des Ball (1980; 1987; 1988).

North West Cape

The establishment of the North West Cape communications station in Western Australia was announced in 1962. It became operational in 1967 and was renamed United States Naval Communication Station Harold E. Holt in 1968 in memory of the late prime minister of Australia. The main function of the radio receiver-transmitter at

North West Cape was to provide communications with United States surface ships and submarines, especially missile-firing submarines, in the Indian and western Pacific oceans. The significance of North West Cape declined in the 1980s as new United States submarines with longer range missiles could be deployed in the eastern Pacific, closer to other United States communication systems. The facility was then passed over to Australian control.

Pine Gap and Nurrungar

Pine Gap is a valley 19 kilometres from Alice Springs in central Australia; an agreement for a United States base in this valley was signed in 1966. Nurrungar is in the Woomera area of South Australia; an agreement for its use was signed in 1969. The facilities became operational in the early 1970s. Pine Gap was under the control of the United States Central Intelligence Agency (this was not revealed until later in the 1970s) and Nurrungar was maintained by the United States Air Force. Several United States agencies have been involved in these bases. The functions of the two bases partly overlap.

Pine Gap and Nurrungar are the ground facilities for a variety of United States satellite systems. They gather and process data from satellites and transmit this information to the USA. Satellites using these ground stations are involved in many different activities. These include:

- photographic and electronic surveillance of other countries
- the monitoring of radio signals from other countries (including all microwave telecommunications within Australia)
- early warning of missile launches
- detection of nuclear explosions
- secret communications by American agents and electronic sensors in different parts of the world.

These two facilities are under United States control.

DEBATING THE ALLIANCE

There has been a great disparity of power between the two parties to the Australian–United States alliance. The USA has, since the Second World War, been the world's strongest military power and largest economy. At best, Australia is a middle power within the international system. This disparity can encourage the view that the alliance was, and is, something foist on Australia by the USA. But the historical record of the origins of ANZUS shows that both Labor and Liberal governments have actively sought alliance with a reluctant USA.

The pressures for a formal alliance, in other words, were as much internal and domestic as external. These internal pressures were based on, or rationalised by, a particular perception of world politics. For some, there may have been an attraction to the idea of a principled commitment to assisting the USA in the containment of communism, and an expression of solidarity with the USA and the West in this cause. But the common perception was of Australia's insecurity and of the need for protection. The most common case for a military alliance was that Australia, with its own resources, could not provide a defence against the military threats that were likely to emerge in the region. At its most basic level, the drive for alliance was motivated by fear: fear of a resurgent Japan at the end of the Second World War; — fear of 'another Japan' in the 1950s and 1960s — of China and communism, of Indonesia, of 'falling dominoes', perhaps even 'fear of Asia'.

The most frequent metaphor for the alliance was that of an insurance policy, by which fears could be alleviated through the securing of 'great power' guarantees of protection, in anticipation of which Australia would pay premiums. The commitments to forward defence and to host United States installations were widely described as insurance premiums, paid in return for a United States guarantee to protect Australia from a possible military threat.

Many Australians have discussed and debated the alliance, although debates and discussions have rarely been initiated by governments. James Richardson states that, in comparison with Britain and the USA, Australia is characterised by an absence of 'great debates' in foreign policy (Richardson 1991). But if there has been a relative absence of debate between the major parties, there has been substantial discussion within social movements and minor parties, and inside the ALP itself.

Three key questions that have been debated over the decades since the alliance was formed are:

1 Does Australia need an alliance 'insurance policy'?
2 Does Australia get reasonable 'cover' from its policy?
3 What does it cost?

1 The 'essential pillar of security': Does Australia need an alliance?

Is an alliance necessary to meet the threats that Australia is likely to face? The answer to this question depends on an assessment of the level of threat Australia faces. In the absence of an 'enemy at the gates', threat assessments are based to a large degree on speculation and conjecture about trends in global politics, and about the intentions and capabilities of other states.

Australian governments in the 1950s and 1960s presented a picture of an Australia threatened by the 'falling dominoes' of communist expansionism. The 'domino theory' was the simple proposition that a communist victory in one state would lead to pressure being put on its neighbour, so that states would 'fall' to communism as rapidly as a series of adjacent dominoes standing on their ends fall over once the first is toppled. In 1950 Spender argued this view, which, fifteen years later, was to underpin involvement in the Vietnam War: 'Should the forces of Communism prevail and Vietnam come under the heel of Communist China, Malaya is in danger of being outflanked and it, together with Thailand, Burma and Indonesia, will become the next direct object of further Communist activities' (House of Representatives, *Debates*, 9 March 1950, p. 627). In 1954 the Minister for Defence, Philip McBride, made a clear link between threat perception and the need for alliance:

> It is a matter of vital importance to maintain the gap between Australia and the present high-water mark of the southward flow of communism. Should this gap narrow, the nature and scale of attack on Australia would become intensified as distance shortened. Finally, should the tide of Communism lap on our shores, we would face an intolerable defence burden and a scale of attack which would be beyond our capacity to repel alone. There is, therefore, every reason ... why Australia should co-operate to keep aggressive Communism within its present boundaries and to stem its onward flow.
>
> (House of Representatives, *Debates*, 28 September 1954, p. 1630)

Dissent from the 'big threat' assessment came from a variety of sources. In Australia communists and their sympathisers dismissed any idea of a communist threat. But leaving these groups aside, the more substantial critique came from 'liberal' academics and former foreign-affairs officials, with books such as *In Fear of China, Communism in Asia: A Threat to*

Australia?, *The Crisis of Loyalty*, and somewhat later, *The Frightened Country* (Clark 1967; Wilkes 1967; Grant 1972; Renouf 1979). These writers argued that the 'domino theory' was based on a flawed understanding of the world. In particular, the theory failed to account for the role of nationalism in the conflict in Vietnam, where communism succeeded largely because it had led the anti-colonial struggle against the French. Many Vietnamese saw the war with the foreign forces of the USA and Australia as a continuation of this nationalist anti-colonial struggle. These writers argued that Vietnamese nationalism would, in turn, constrain China's influence over Vietnam. They also observed that Australia had a history of exaggerating threat, which dated from the nineteenth century, when the colonists periodically experienced panics about attack or military pressure from France, Russia, or Germany. This 'threat mentality' is seen as an understandable development in a British settler colony located far from the 'mother country'. Cultural and economic isolation from the Asia-Pacific region contributed to a fear of Asia in the twentieth century.

The 'low threat' view gained a mass audience through the widespread political mobilisation against fighting the Vietnam War and military conscription that occurred in the late 1960s and early 1970s. When the USA itself sought rapprochement with China, the idea of falling dominoes had become passé. With the election of the Whitlam government, the low threat outlook became the official outlook. Whitlam's minister for defence, Lance Barnard, stated that Australia would face 'no threat for fifteen years', and while this shocked the domino school, it proved accurate.

The rhetoric of Malcolm Fraser, in the second half of the 1970s, returned briefly to the 'big threat' imagery, initially in concern expressed over Soviet naval expansion in the Indian Ocean, and then with the Soviet invasion of Afghanistan.

With the return of the ALP in 1983, the low threat outlook became entrenched in defence policy-planning documents. The 1986 report to the Minister for Defence by Paul Dibb began with the words:

> Australia is one of the most secure countries in the world, it is distant from the main centres of global military confrontation, and is surrounded by large expanses of water which make it difficult to attack. Australia's neighbours possess only limited capabilities to project military power against it … Australia faces no identifiable direct military threat and there is every prospect that our favourable security circumstances will continue.

(Dibb 1986)

The question arises — why does a country so secure need an alliance?

In the low threat outlooks of Labor governments in particular, the need for alliance would appear to be radically diminished. But Labor governments did not take the 'logical' step and canvassed the issue of the need for the alliance. Perhaps a bedrock of realist assumptions remained: in an uncertain world, why not keep on an insurance policy, even if no threats are now apparent? An alliance may not be an essential and necessary relationship, but it may be an additional and useful deterrent.

The realist case for continuing with a possibly useful insurance policy may be persuasive, good 'common sense'. But as with all such policies, we can ask 'what is the cover provided, and what is the cost?'. Ironically it is the realist perspective that then generates a deep scepticism about the official answers to these questions. Insurance cover may be worth having, but can you get it?

2 'The mighty shield against any armed attack': Does Australia get protection?

The logic of realism, which supports retaining an alliance for its additional insurance against an uncertain future, also provides the strongest critique of the capacity of an alliance to deliver what is sought or expected. In a pamphlet titled *Australia: Armed and Neutral?* written in the mid-1960s, academic Max Teichmann (1966) articulated a realist critique of alliance that has reverberated in the debate ever since: whatever the case may be for pursuing an alliance, the kind of commitments that Australia was seeking from an ally would, he argued, always remain elusive. The realist proposition is straightforward: great powers act according to their perceived interests, which are calculated from a global perspective. Whether a great power will meet the obligations of its treaties will depend on the calculus of interest at the time. There are no guarantees in an anarchic system.

This argument resonates in almost all critiques of the alliance. Three historical moments in the history of Australian foreign policy underline the point: the events surrounding the signing of the ANZUS treaty itself; the United States response to the Indonesian incorporation of West New Guinea; and the Guam (or Nixon) Doctrine of 1969.

The ANZUS treaty was primarily a result of Australian–United States bargaining. The USA agreed to a treaty rather than a presidential declaration in order to gain Australia's support for the 'soft' peace treaty it was proposing to conclude with Japan. In turn, Australia agreed to a pact that was less strongly worded than the one that established NATO. As Tom Millar has written, ANZUS 'was not sought by the United States, which acquiesced only with reluctance' (Millar 1978, p. 207).

Article III of the treaty requires the USA to 'act to meet the common danger according to its constitutional processes', and these constitutional processes might allow the United States Congress to overrule the administration (or president). But the United States administration was not keen to have itself bound by Australia in any case. In 1951, John Foster Dulles, the United States diplomat primarily responsible for the ANZUS negotiations, wrote to General Douglas Macarthur in a letter made public in the 1970s. Dulles stated that 'While [ANZUS] commits each party to take action, it does not commit any nation to action in any particular part of the world. In other words, the United States can discharge its action against the common enemy in any way and in any area that it sees fit' (Foreign Relations of the United States 1951, p. 177).

In the case of West New Guinea, Australia unsuccessfully opposed Indonesia's efforts in the 1950s and early 1960s to incorporate this territory (the remaining part of the Dutch East Indies) into the new state of Indonesia. Australia sought United States support for its position but did not receive it. The USA decided that its interests were better served if Indonesia were allowed to incorporate the area, and Indonesia acquired 'West Irian', later renamed 'Irian Jaya', in 1963. (Two years later, when Australia gave military assistance to Britain against Indonesian 'Confrontation' with Malaysia, the USA refused to make any public statement that it would be obliged by ANZUS to support Australia.)

The Guam (or Nixon) Doctrine of 1969 refers to the statement of President Nixon on the United States Territory of Guam in the Pacific in the wake of the rising casualties and intractability of the Vietnam War. The message of Nixon's Guam Doctrine was that the USA's allies, irrespective of what was written in their treaties with the USA, were expected to take a much greater responsibility for their own military security, unless threatened by major powers involving nuclear weapons. The USA expressed an unwillingness to commit ground forces to the Asian mainland and a general reluctance to become directly involved in military conflicts in the Asia-Pacific area and elsewhere.

Against the logical force of realist scepticism about alliances, those wishing to believe in the United States 'shield' presented arguments from rationalist and even idealist standpoints.

The rationalist argument stressed that the USA had made public commitments under ANZUS, which it would be likely to meet to retain credibility with other allies, including Japan and the NATO powers, with whom the USA also had security treaties. A failure to honour the ANZUS treaty would severely weaken other treaties that were even more important for United States interests. But this argument is reminiscent of the early days of the cold war, when the USA sought an interlocking series of alliances to contain the USSR and China.

The idealist argument was that Australia had a special relationship with the USA: that Australia had a deeper tie than just an insurance policy. The idea of special relationship, however, was rather vague. It could refer to shared Second World War experiences, to shared political identity as part of the liberal-democratic-capitalist West, or perhaps it could hint at the idea of a White, English-speaking identity. In other words, the United States–Australian relationship had some of the qualities that once existed in Australia's relationship with Britain. The idea of a special relationship had more currency in some Australian conservative circles than it did in the USA, and today it does not seem to have any currency in either (Phillips 1988).

We are left with the 'double negative' outlined in the 1987 White Paper on defence: that an adversary cannot be sure that the USA will not support Australia, and that this is still better than no alliance. If this is a gamble rather than a guarantee, what are the odds, and how much should be put on the table by Australia?

3   An acceptable cost?

There is no debate about whether the policy of insuring through alliance has had costs for Australia and Australians. The most obvious cost was the deaths of Australians in the Korean and Vietnam wars, and the physical and emotional trauma suffered by troops sent to these wars and by their families. The question has been whether the price of 'forward defence' was worth paying. For the Liberal-led coalition governments of the 1950s and 1960s, the 'domino theory' and the idea of the United States shield led to this question being answered in the affirmative. But when the Vietnam War turned into a quagmire, the Labor Opposition shed its ambivalence and answered 'no'. The Whitlam government rejected the concept of forward defence, but any potential conflict with the USA over this rejection was minimised by the fact that the USA was implementing its withdrawal from Vietnam at this time.

The establishment of United States installations in Australia was initially surrounded by 'secrecy, evasion and deception' on the part of Australian governments (Ball 1980). Arguments about the pros and cons have moved through several phases. Critics were initially concerned about the loss of sovereignty inherent in United States-controlled facilities, and about the lack of information on what the bases really did. Liberal governments were reluctant to admit the military functions of bases to the Australian public. When acknowledging the military role of these facilities, governments reiterated that the bases reinforced the United States commitment to Australia and were part of Australia's contribution to ANZUS under Article II.

A new realist argument for the alliance was suggested: that the USA now had a stake in Australia's protection as Australia was host to such vital installations. There was also a realist reply: that the USA's protection of its bases was not the same as the protection of Australia, and as long as any adversary of Australia's respected United States interests, the USA may not intervene. Some supporters of the installations argued that they gave Australia leverage over the USA in a crisis, but this was not an argument that was put by governments, as the idea of pressuring an ally to act did not sit well with the idea that the ally would act in any case out of treaty commitments.

With the reintensification of cold war antagonisms in the early 1980s, criticism of United States bases focused on their role as potential Soviet nuclear targets, as Pine Gap and Nurrungar would play crucial roles in any United States nuclear exchange with the USSR. Striking the 'eyes and ears' of the United States military early in such a war could be seen as a high priority for the USSR. Furthermore, some of the functions exercised by United States bases were argued to be destabilising as they made the USA capable of launching a first strike (Ball 1987). These arguments were articulated in Australia by the burgeoning disarmament movements of the early and mid-1980s. The 'bases debate' became one between the peace movement and the Hawke Labor governments. Australia's foremost expert on United States installations and on United States nuclear strategy, Des Ball, came down somewhere in the middle, publishing a detailed case for the closure of the facility at Nurrungar, although supporting the continuation of Pine Gap.

In reply, the Labor government argued not on the traditional grounds that 'joint facilities' would shore up the United States security guarantee, but on a new tack: that the facilities contributed to global nuclear stability and to arms control efforts through their capacity to monitor the Soviet military situation.

The debate on United States installations was conducted from these intellectual trenches until the end of the cold war demolished the assumptions of both sides. United States bases ceased to be targeted; superpower nuclear arms reductions took place without any necessary reliance on these facilities. Nurrungar found new support from former critics on the basis of its potential role in monitoring nuclear non-proliferation in the region. Pine Gap remained shrouded in secrecy (Brown 1994).

The lines of a post-cold war debate are not yet drawn. The sovereignty issue raised at the start remains. The USA has a series of facilities that it utilises in pursuing the military dimensions of its foreign policy. How compromised is the credibility of an independent Australian foreign policy when, no matter whether Australia agrees or disagrees with the United States stance on an international issue, or whatever position it takes in the UN and other international forums, United States facilities in Australia will be used to support the United States foreign-policy stance?

THE ALLIANCE AS FOREIGN POLICY: HOLDING THE LINE

Under the leadership of a Liberal–Country Party government in the early 1950s, Australia chose to enter an alliance with the USA. Theoretically, other choices were possible, but the government gave no consideration to non-alignment or to regionalism based on non-alignment. Liberal governments for twenty-three years accepted the insurance logic and paid the premiums sought by the USA: forward defence commitments and hosting United States installations

The Whitlam government and the alliance

The ALP, in Opposition for twenty-three years, contained the whole range of views within it, from strongly pro-alliance positions (usually associated with the Right) through to those sections of the Left that wished to pursue a non-aligned path and leave the United States alliance. The leader of the Labor Opposition, Dr Herbert Vere Evatt, and most of the immediate postwar ALP, supported the signing of ANZUS in 1950 and reminded Menzies that they had initiated the quest for such a treaty. Unease in the ALP was generated not by the ANZUS treaty, which seemed to offer something to Australia, but by the intensification of the cold war and the 'premiums' that were paid by Australia in the 1950s and 1960s. However, as the premiums increased, so did the influence and numbers of those in the party who questioned the treaty itself. The 1955 split caused part of the Right to defect to the newly formed DLP and diminished the strength of pro-alliance sentiment in the remaining party. But a non-aligned position never secured the endorsement of the party at a national level, although it became part of the official platform of the Victorian branch. The federal parliamentary leadership of the ALP successfully resisted those on the Left who wanted to break the alliance with the USA. They argued that an independent foreign policy was not incompatible with an alliance relationship and that Australia could achieve more independence within the alliance than Liberal governments had pursued. However, the federal party platform did express opposition to foreign bases.

When the ALP finally achieved office in 1972, 'a more independent stance in international affairs' was a major theme of Whitlam's foreign-policy outlook. Whitlam created a sense of rapid change with a series of foreign-policy initiatives in the first days of government. But with the Vietnam War raging, passions within the ALP took this independence further than Whitlam wished to go. In December 1972, when United States President Nixon ordered the resumption of heavy bombing of North Vietnam, senior Labor ministers Tom Uren and Jim Cairns described United States actions as 'mass murder' and 'the most brutal and indiscriminate slaying of men and women in living memory'. The Nixon administration objected strongly to these attacks, and to bans imposed by Australian maritime unions on United States shipping to South Vietnam.

In response to the USA's displeasure, Whitlam stressed that the USA should not take it for granted that Australia would necessarily agree with it on all matters of common concern. However, he also tried to reduce the potential for conflict with the USA over the Vietnam War issue by showing restraint in official public statements.

He rebuked other Labor ministers for speaking outside their minister-ial areas and for attacking United States policies. The particular issue was partly defused by the cessation of United States bombing and the signing of the Paris Peace Accords on Vietnam in January 1973. For the duration of the Vietnam War the Labor government emphasised its 'even-handed' criticism of both sides and remained largely silent on other United States bombing attacks.

The Labor government also modified its approach to SEATO, largely in deference to United States wishes. When he was in Opposition, Whitlam had described SEATO as moribund and an anachronism from the period of the containment of China. However, when in government, Whitlam did not pull Australia out of the treaty organisation but supported moves to gradually phase it out.

Although the ALP's platform opposed the presence of foreign-owned, controlled, and operated bases in Australia in peacetime, the Whitlam Labor government was very cautious. It removed none of the secrecy from Pine Gap and Nurrungar, and did not seek to renegotiate the agreements for these bases, asserting that they could not be used to make war on other countries because they were not involved in the command or control of nuclear weapons. Whitlam pursued a minor renegotiation of the North West Cape agreement, concluded in 1974.

To sum up, the ALP's ambivalence towards the alliance and its desire for 'more independence' did not, under Whitlam's leadership, translate to a challenge to any aspect of the alliance relationship. Withdrawal from Vietnam was United States policy, as it was Australia's, and the ALP did not seek to alter the arrangement for United States bases in Australia, except in a token way.

The Fraser years

After the coalition had been elected in its own right in December 1975, the new Prime Minister, Malcolm Fraser, announced a new emphasis and approach in foreign policy, the main elements of which were a more pessimistic view of international trends, both in the region and in the world as a whole; expressions of alarm and concern regarding a perceived increase in Soviet naval activity in the Indian Ocean, and regarding Soviet policies generally; and a reiteration of the importance of the United States alliance.

Fraser set out to extend Australia's military cooperation with the USA, especially in areas affecting the Indian Ocean. He lifted the ban on the entry of nuclear-powered naval vessels into Australian ports. It had been imposed by the McMahon Liberal government for safety reasons and retained by the Whitlam government. Subsequently, United States nuclear vessels visited Australian ports. The Fraser government immediately accepted the USA's request for a further ten-year agreement on Pine Gap, rather than the annual leasing arrangement that had existed after the first ten-year agreement had expired. It also acceded to the establishment of a new United States satellite ground station at North West Cape. Australia offered expanded use of the Cockburn Sound naval facilities and Learmonth air base, both in Western Australia.

Fraser's desire to reinforce the alliance was a direct continuation of the traditional Liberal approach of earlier decades. The key difference was that it focused solely on the 'Soviet threat', completely rejecting the 'China threat' and the idea of falling dominoes under China's influence. Fraser even went as far as to applaud a Chinese missile test in the Pacific.

The ALP under Hawke: the peace movement and the ANZUS crisis

The Labor Opposition found itself uneasy with some of Fraser's moves to increase military commitments to the USA, although it tended to avoid outright rejection of government initiatives. Bill Hayden, who had replaced Whitlam after the 1977 election loss, opposed landing rights for *nuclear-armed* B-52 United States bombers rather than all landing rights. Hayden also opposed the use of Cockburn Sound as a *home* base for United States warships, rather than its role as any kind of United States base. Some more substantial differences on the Australian–United States alliance began to emerge in the early 1980s after Ronald Reagan was elected United States President and cold war tensions returned. Hayden drew increasing attention to the costs of United States bases, noting the danger that they represented as nuclear targets. In 1981 he returned to the ALP's traditional concerns about the North West Cape and sought further renegotiation for effective Australian control.

These sentiments would soon receive considerable support with the emergence of a mass peace movement in Australia, paralleling developments in Western Europe. The Australian peace movement was a diverse coalition, but it was generally united in the desire to 'close United States bases' in Australia because of their connection to the nuclear arms race, and to ban the entry of nuclear-powered or armed vessels into Australian ports. The peace movement had considerable impact on the ALP as many branches took the issues to the party leadership and sought to incorporate them in the party platform.

Bob Hawke toppled Bill Hayden as Labor leader and led the ALP to victory in the 1983 election, and to subsequent wins throughout the 1980s. Hayden became foreign minister in the new government. Hawke was determined not to implement the party platform on United States bases and mobilised to have the platform changed at the party's 1984 National Conference. Disaffected groups formed the Nuclear Disarmament Party (NDP), which then joined the Australian Democrats in advocating the closure of United States bases in Australia, an end to visits by nuclear-armed and powered ships, and an end to the use of Australian airports for B-52 training flights. The NDP had some success at the 1984 election and joined the Democrats in holding the balance of power in the Senate.

Hawke's 1983 visit to the USA was largely to reassure the United States government that he had not 'bought' the peace movement agenda. On the other hand, the ALP sought to woo the 'peace vote' through a series of initiatives: the appointment of an ambassador for disarmament, the establishment of a Peace Research Centre in Canberra, support for the International Year of Peace in 1986, and a more active pursuit of international arms-control matters, particularly a comprehensive test ban treaty and a South Pacific nuclear-free zone. Hawke and Hayden claimed to be pursuing arms control and disarmament issues in ways compatible with the alliance, and argued that being the USA's ally would give them additional influence over these matters.

The Hawke government opposed the USA on one issue of principle: the multi-billion dollar 'Star Wars' project, or 'Strategic Defence Initiative', in which Australia refused to participate because of what were seen as the destabilising implications of the proposed scheme. The Hawke government took a strong declaratory stand against the 'Star Wars' proposal of 1983 and declined the subsequent United States offer to participate in research. A familiar issue arose: what if the United States bases in Australia were used in ways that contributed to a project opposed by the Australian government? The Australian government claimed that the United States bases would not be involved.

However, Ball cogently argued that Nurrungar would be an integral part of the project for as long as it remained part of the United States early-warning system (Ball 1987).

Hawke was overruled on one specific matter: cooperation with the USA on the 'MX' missile project. In early 1985, while Hawke was visiting the USA, the Australian press revealed that the United States government had sought Australian support facilities for the proposed testing of MX missiles, which would splash down in the Tasman Sea. Furthermore, a small group of Cabinet ministers had secretly agreed to continue the approval that the Fraser government had granted. There was such a strong reaction in the ALP to both the decision and the secrecy surrounding it that Hawke's advisers predicted he would lose the issue in the Caucus when he returned to Australia. He announced while overseas that Australia would not offer support facilities after all. The USA glossed over the matter, partly because the MX project had an uncertain future in any case, and partly because of its greater preoccupation with New Zealand and the ANZUS crisis.

The ANZUS crisis

The period from mid-1984 to mid-1986 has been described as the 'ANZUS crisis'. The 'crisis' arose from the conflict between New Zealand and the USA over New Zealand's ban on the entry of nuclear-armed and/or powered warships. Australia was presented with the problem of responding to the conflict between two allies in a way that preserved good relations with both and yet retained Australia's alliance with the USA.

The Labour government of David Lange was elected in New Zealand in mid-1984. Part of the New Zealand Labour Party's policy platform was the port ban. United States warships visited New Zealand infrequently, and there was no immediate military significance in Lange's move. However, the USA argued that its ability to offset Soviet penetration in the Pacific would be hampered severely by such port bans, especially if other countries followed New Zealand's lead. The Soviets, it argued, had expanded their Pacific fleet considerably, and port facilities were essential for the USA to provide a credible deterrent to the USSR. For strategic reasons, the USA would neither confirm nor deny the presence of nuclear weapons on any specific ship. In effect, this meant a New Zealand ban on all United States warships.

The USA maintained that such port visits were an essential element of cooperation under the ANZUS alliance. The New Zealand government stated that it wished to maintain the ANZUS alliance but to exclude all nuclear cooperation, arguing that such cooperation was not part of the original ANZUS treaty.

The Australian (and United States) governments initially believed that New Zealand would simply back down under pressure and adopt the same pragmatic approach as the Australian Labor government had towards United States bases. The view failed to take into account the widespread support in New Zealand for the policy. In the election that saw the Labour Party come to office, the New Zealand Party, on the Right of the political spectrum, had campaigned on a more radical platform than the Labour Party: total withdrawal from the ANZUS alliance. New Zealand stuck with its port bans and incorporated the decision into legislation.

The USA reacted by ending military and intelligence cooperation with New Zealand. In 1986, the USA declared that it did not consider itself bound by ANZUS to come to New Zealand's assistance against armed attack. Some United States politicians threatened trade retaliation as well.

After initial criticism of the New Zealand stand, the Australian government's position developed along two lines. First, Australia opposed United States moves for a new bilateral treaty to replace ANZUS. The Hawke government sought to retain the ANZUS treaty as the formal document. It did this partly in the hope that New Zealand might eventually change its policy, and partly to forestall moves to have Australia sign a treaty that was more specific than ANZUS, which could impose greater obligations. Australia assured the USA that it would not follow the New Zealand example. As a consequence, the ANZUS treaty remains, even though the USA has stated that it does not operate as far as United States–New Zealand relations are concerned.

Second, the Hawke government did not join the USA in cutting intelligence and military links with New Zealand, and maintained Australia's commitment to assist New Zealand under the ANZUS agreement. While Australia criticised New Zealand, it did not follow United States retaliatory moves.

The 1986 superpower summit at Reykjavik in Iceland was the first substantial easing of cold war tension since the late 1970s, and it fore-

shadowed the vigorous *détente* that developed under Gorbachev, leading to the end of the cold war and then to the end of the USSR.
Disarmament movements declined rapidly as the fear of war receded
and superpower arms negotiations continued. The domestic pressure
on the Hawke government from the peace movement eased and then
evaporated.

The Hawke government had responded to the pressure from the
peace movement by pursuing disarmament initiatives that were not seen
to be in conflict with the central features of the United States alliance. As
with the Whitlam government in the 1970s, it sought to redirect and
limit those elements of the party who sought to challenge the alliance.
The New Zealand Labour Party, by contrast, embraced an element of
the peace movement agenda and found itself in sharp conflict with the
USA over ANZUS. The Australian government sought to preserve the
ANZUS treaty despite the New Zealand–United States conflict.

CONCLUSION: ALLIANCE IN DECLINE?

These were conflicts generated by the cold war, and the end of the cold
war removed them from the agenda. In the period lasting from
Whitlam to Fraser to Hawke, from 1972 to the end of the cold war in
the late 1980s, bipartisanship on the alliance was achieved through the
ALP controlling its Left wing, and losing a section of support to the
NDP and independents. The Liberal Party also lost a section of more
critical opinion on its Left with the creation of the Australian
Democrats. The idea of alternatives to alliance was promoted outside
the major parties by Democrat, Nuclear Disarmament and Green sen-
ators and independents. In parliamentary terms, the fundamental
debates on the alliance were confined to the Senate.

In the post-cold war years, although bipartisanship may continue
to be evident on the issue of the United States alliance, the significance of that alliance to Australian foreign policy seems to have substantially declined.

The ANZUS crisis, precipitated in 1984 by New Zealand's ban on
the entry of nuclear-armed and powered vessels, had a very significant
consequence for Australia: it robbed the Australian–United States
alliance of its traditional name, 'ANZUS alliance' — a term that identified the treaty at the core of the relationship and that had a similar
ring to NATO alliance. The annual ANZUS Council meeting

disappeared and was replaced by the Australian–United States Ministerial Talks. The meeting had become mere 'talks'; and the new acronym, AUSMIN, failed to enter the political discourse.

The loss of name paralleled a more widespread sense of the declining significance of the alliance to Australia's military security. The 1969 Guam Doctrine had unilaterally cut United States commitments to all its allies, and the 1986 Dibb Report and 1987 White Paper on Australia's defence can be seen as the first serious Australian attempts to plan for the self-reliant defence of Australia, without assuming that 'Uncle Sam' will come to Australia's aid. The report challenged the entrenched elements of forward-defence thinking that remained in the military, and redirected spending priorities more towards the direct defence of Australian territory and its maritime approaches.

The Dibb Report did not propose a break with the USA, and the concept of self-reliance was fudged to accommodate continuing dependence on the USA for intelligence, resupply, and spare parts. But the concept of self-reliance was also intrinsically in tension with the idea of alliance. The Foreign Minister, Evans, declared that the new defence policy had 'liberated Australian foreign policy'. What he clearly meant was that Australia could assume a much greater measure of independence in foreign policy as it was making itself less dependent on the idea of United States military protection. Evans reminded Australians that 'good relations' with the USA were not an end in themselves but a means of securing Australian interests, and that policies needed to be assessed by this yardstick.

The idea that ANZUS was 'the cornerstone' of Australia's foreign policy underwent a subtle but fundamental shift: ANZUS ceased to be described as 'the cornerstone' and became simply 'a cornerstone'. The first kind of cornerstone is that 'built into a corner of the foundation of an important edifice as the actual or nominal starting point in building, usually laid with formal ceremonies ... and made the repository of documents' (*Macquarie Dictionary*). This is a very rich and suggestive metaphor for the early years of the ANZUS treaty, and by definition, there could only be one such cornerstone. The other kind of cornerstone is that which 'lies at the corner of two walls and serves to unite them'. There can be many of these cornerstones depending on the height of the walls and the number of corners.

As Australia's military security expectations of the United States steadily declined, its economic expectations were increasingly frustrated. Hawke's 'bullet in the head' outburst about the United States Export Enhancement Program (EEP) signalled a deep and continuing level of frustration with United States policies, a frustration that extended into the 'conservative heartland' that had traditionally supported the United States alliance. West Australian farmer organisations argued that Australia should use United States bases as a lever in negotiations with the USA by threatening to close the bases unless the USA changed its agricultural subsidy policy.

The United States alliance has not been declared defunct. Its steady marginalisation with regard to Australia's military security concerns and its inability to assist economic security concerns have not led to an overall attempt to redefine and re-evaluate the relationship with the USA. Instead, concepts that attempt to locate Australia in the world without reference to the old identification of Australia as ANZUS ally have increasingly been propounded. These are the ideas that:

1 Australia is part of the region and formulates foreign policy in association with neighbours, through economic involvement in APEC and participation in the ASEAN regional security dialogue, and

2 Australia is a middle power on the world stage, taking foreign-policy initiatives in association with like-minded powers with shared interests, wherever those powers may be.

We now turn to these other cornerstones of foreign policy.

4 Australia as Part of the Region

Australia's geography, it seems, is now its destiny. With the end of the cold war, Australia has rediscovered the importance of 'region'. There is a sense that its engagement with the region has been delayed against its best interests, partly because of the European origins of more than 90 per cent of the Australian population. Gareth Evans developed this theme throughout his long tenure as foreign minister:

> The great turn around in contemporary Australian history is that the region from which we sought in the past to protect ourselves — whether by esoteric dictation test for would-be immigrants, or tariffs, or alliances with the distant great and powerful — is now the region which offers Australia the most. Our future lies, inevitably, in the Asia Pacific region. This is where we live, must survive strategically and economically, and find a place and role if we are to develop our full potential as a nation.
>
> (Evans & Grant 1995, p. 348)

The 'turn around' that the foreign minister sought both to describe and pursue followed two centuries of postponement that was permitted by the continuing British connection and extended by the postwar American connection. But economic and strategic developments have pushed the idea of region from the margins towards the centre of Australian foreign policy.

GLOBAL CONSTRUCT 1: EMPIRE

Throughout the period of Australia's colonial settlement in the nineteenth century, and continuing after federation through until the Second World War, the organising principle of Australia's external relations was 'Empire', the British Empire. The British Empire was a global political system, the components of which were coloured red on the maps used by British and Australian school children. Identification with Empire came naturally to many Australians. Britain was itself the source of almost all the immigrants to Australia, and Australia's trade and investment links were focused on the 'mother country'. Britain was a world power of the first rank. Australian governments after federation assumed Britain to be responsible for Australia's security. Australian foreign policy until the Second World War, such as it was, was largely confined to attempts by a loyal dominion of the Crown to influence British government positions on particular issues. There was no Australian diplomatic representation in the region.

Distance from Britain may have been a 'tyranny' (Blainey 1966) which ruled key aspects of Australian economic, social, and political life, but freedom from such a tyranny was not sought in relationships with states and societies closer by. The region was something that Australians and their governments looked across and sailed past to Britain, the place that generations of Australians persisted in calling 'home' when they, and even their parents, may have been Australian-born. If the region mattered at all, it was largely, as Gareth Evans suggested, as a source of anxiety.

From the mid-nineteenth century, China, or rather the Chinese people, were seen by the settlers in Australia as the 'Yellow Peril' — as unwanted immigrants who came to Australia, like many others, in search of gold. Australian colonial parliaments moved to stop Chinese immigration, and well before federation in 1901, the 'White Australia' policy was in place, prohibiting immigration from almost everywhere except Britain — from most of Europe and from all of Asia.

This betrayed a cultural, rather than a military, insecurity on the part of British-Australians, who had made 'race' central to their identity. The colonists' military anxieties in the nineteenth century were not focused on Asia but on those European states that, at different times, were Britain's adversaries or rivals: France, Germany, and Russia. In the second half of the nineteenth century, colonial governments built some attractive fortifications at the entrances to ports and harbours, and expressed concern to Britain about the possibility of its imperial rivals establishing outposts in the Pacific islands. Some Australian colonies agitated for pre-emptive British colonisation of island groups.

The Queensland government even attempted to annex part of New Guinea in the 1880s to forestall Germany, and although this move was rejected at the time by Britain as being outside Queensland's competence, Britain later annexed part of New Guinea and then passed control to the new federal government of Australia. The South West Pacific islands were that part of the region with which Australians had some substantial engagement in the nineteenth and early twentieth centuries. The islands were used as an economic hinterland for the settlers of the east coast of Australia — as a source of agricultural and ocean resources, and a supply of cheap or semi-slave labour for sugar cane plantations.

In the early decades of the twentieth century, fear of Britain's European rivals in the region subsided. France fought together with Britain in the First World War. German colonies in the Pacific islands — in New Guinea, Nauru, and Micronesia — were seized by Australia and Japan at the outset of hostilities and not returned after the war. The Russians had previously been dealt a devastating military blow by Japan in a war in 1905, and in 1917 Russia became convulsed by revolution.

As anxiety over European military threats dissipated in the first decades of the twentieth century, it was replaced by anxiety over Japan — the first regional power to be seen as a potential military threat. Prior to the outbreak of the Second World War, the Prime Minister, Menzies, expressed the view that what Britain viewed as the 'Far East' was, for Australia, the 'Near North'. What might appear today to be a statement of the obvious was seen at the time as an astute observation, acknowledging that Britain might not accord the same priority to East Asia as Australia was required to do by geographical proximity. No action followed from Menzies's observation about the tension between British and Australian perceptions. When Japan was transformed from potential to actual threat in December 1941, substantial Australian forces were already deployed overseas in support of Britain's wars in Europe and the Middle East. Former diplomat Malcolm Booker, in *The Last Domino* (1977), put the case that the Australian commitment to Empire led it to be largely unprepared for defence against attack from other forces in the region. The point was not that Britain 'let us down', but that Australia had not set its own priorities to take account of the regional context, and relied too heavily on Britain.

The British Empire was effectively destroyed by the Second World War, and Australia lost the global power structure in which it had posi-

tioned itself, and that provided its international identity and status.
The imperial connection had allowed Australia to approach the region
with a mixture of ignorance, indifference, and apprehension.

GLOBAL CONSTRUCT 2: ALLIANCE

The United States alliance in the 1950s and 1960s offered Australia a
new global structure: 'West' versus 'East', in which the leading
Western powers were the USA and Britain, and the leading powers of
the East were the USSR and China. Australia aligned itself with the
West and approached the region through the ideological prism of the
cold war — a prism that refracted the complex spectrum of the region
into black and white:

- Communist Asian states and movements were assumed to be
  'bad', regardless of the political history of particular communist
  movements and the strength of the nationalist credentials in their
  societies. The 'Yellow Peril' was now the 'Red Menace' as well. On
  many maps red no longer indicated the British Empire but was
  used to depict communist states.
- Anti-communist Asian states were assumed to be 'good' states, *Suharto's Indonesia*
  regardless of the level of internal political repression or lack of
  commitment to liberal democratic practices.
- Non-aligned states were, actually or potentially, dupes of the communists, likely to succumb to internal and external pressure from
  them.

Through its commitment to alliance, Australia became engaged in
the region as the ally or adversary of states and peoples whom it hardly knew, reacting to their anti-communist or communist ideology.
South Vietnam was classified as an ally, North Vietnam as an adversary, and Australians set off to kill and be killed in a place most of
them had not heard of. Japan, on the other hand, despite the bitter
memories, was reclassified as 'West' in the East–West conflict. As
allies of the USA, Australia and Japan developed a fast-growing economic connection, with Australia providing raw materials for
Japanese industrialisation. The 1957 trade agreement between
Australia and Japan established the politico-legal basis on which the
economic relationship could be built. China was the principal adversary (although one with which Australia was prepared to trade) until

United States President Richard Nixon called an end to the cold war in Asia in 1972 and sought to normalise relations.

By making the United States alliance the 'cornerstone' of foreign policy, the Liberal-led coalition of the 1950s and 1960s reduced regional diplomacy to a subordinate role. Once Australia had committed itself to an alliance with a distant great power, what started as an internally driven policy (arising from the need for security felt by Australians) translated into a tight external constraint as the costs of compliance rose; and yet the costs of extrication seemed higher still. Joseph Camilleri has described the pattern of relationships as a 'web of dependence' (1980). To what extent did the alliance create a structure through which the USA was able to assert that its interests were Australia's interests? To what extent could the USA press for particular policies to be adopted by Australia's political leadership, policies that did not necessarily have widespread support in Australia or reflect its developing interests in the region? Camilleri gave the example of pressure from United States companies to develop and export Australia's uranium reserves. Public opinion polls often showed strong support for ANZUS, but a significant degree of opposition to United States bases (Campbell 1989). The secrecy and misinformation surrounding the establishment of United States facilities suggest that political parties lacked confidence that they had public opinion on their side.

In contrast to the 'web of dependence', Philip Bell and Roger Bell argue that Australians have remained 'implicated' in negotiating the terms of their relationship with the USA and have possessed a substantial measure of power to modify United States interests to accommodate Australian concerns (Bell & Bell 1993).

Yet there remains the issue of when such assertiveness has been exercised in foreign policy. What latitude does a junior partner have to disagree with a great power ally without voiding the 'guarantee' that it hopes has been given? The disparity of power between two allied states, A and B, does not necessarily mean that A dominates B. But the disparity in the states' perceptions of the alliance's significance may mean that B develops a psychology of dependence, as the stakes seem much higher for B than for A. Australia may have assumed its alliance to be the 'essential pillar' of military security, but for the USA, its alliance with Australia has been one of many such relationships with weaker powers, relationships that are not at all essential to United States security. The question of whether the USA would deliver protection when it is needed is not just a matter for debate; it is also a condition of anxiety experienced by governments who know in their realist hearts that the idea of a 'guarantee' is fanciful. The anxiety created by this disparity of perceptions may have led to an excessive dependence, and to attempts to ingratiate Australia with the USA.

In the 1950s a belief in the viability of that other 'powerful friend', Britain, created psychological space that allowed a degree of criticism of the USA. But in the 1960s, Holt's slogan 'All the way with LBJ' encapsulated the comprehensive diplomatic dependence to which Australia seemed to aspire. Even Coral Bell (1988), a firm supporter of the United States alliance, characterised the 1960s as a period of excessive dependency. Such a dependency created a vicarious foreign policy in the region, in which Australia followed and identified with United States might and power. It also created a rigidity in policy that made it very difficult for Australia to adapt to the dramatic shift in United States policy that occurred in the early 1970s.

Labor governments have argued that they could pursue 'independence within the alliance', and this formulation has also been adopted by the Liberal-led coalition. It is a central part of the 'bipartisanship' displayed since 1972. 'Independence' suggests foreign policies based on an Australian view of what is in its best interests, with an eye to regional complexities and opportunities; 'within the alliance' suggests foreign policies constrained by what is acceptable to the USA. The potential for conflict is obvious. How have such conflicts been manifest and managed?

In the Whitlam years, a credibility gap emerged in several areas between 'independence' rhetoric, focused on regional initiatives, and United States alliance commitments. When the gap became too obvious, the rhetoric was toned down or undercut by private diplomacy, and regional policy was brought into line with United States interests.

This was the fate of the ALP's declared support for proposals for a *zone of peace* in the Indian Ocean and a *nuclear weapons free zone* in the South Pacific. The concept of a South Pacific nuclear weapons free zone was raised by New Zealand and Fiji in 1975. Australian support was sought for a UN resolution endorsing the idea. Whitlam had supported similar ideas in the United Nations in 1974. He commented that they deserved Australia's support because they addressed the dangers that lay in an uncontrolled and unregulated deployment of nuclear weapons. However, leaked documents from the period have revealed the gulf between words and deeds (see Walsh & Munster 1980). The United States Embassy sent the Australian government a note in June 1975 setting out United States objections to the concept, fearing that the zone would restrict United States naval and air passage, or port calls and use of facilities in the Indian Ocean. This, it was said, affected United States strategic considerations more than those of the USSR or China. The note also stated that the USA could not agree to any proposal that restricted the right to cross the high seas, nor would it confirm or deny which vessels were carrying nuclear weapons.

Subsequently, Whitlam wrote to the Labour Prime Minister of New Zealand, Norman Kirk, reiterating these very points and stating that Australia would not co-sponsor the motion, lobby for it, or even speak in favour of it. Whitlam stated that a limited initial proposal could stir up controversy with the USA and raise questions about the ANZUS relationships. Kirk replied that he thought that their relationships with the USA could accommodate the proposal without any undue strain. In response to this, Whitlam simply restated Australia's misgivings (Walsh & Munster 1980). In this incident lay the seeds of the ANZUS crisis, which were to sprout a decade later.

The Hawke governments in the 1980s found themselves in a similar dilemma: of reconciling strong support for the alliance with a desire to make 'independent' foreign policy in the region. In response to domestic pressures to take a more active stance on nuclear disarmament, the ALP returned to the idea of a nuclear free zone in the South Pacific and set out to establish the South Pacific Nuclear Free Zone Treaty. To accommodate the long-standing United States objections that such a zone might affect naval access through the South Pacific, the Australian government developed a treaty proposal that banned the possession or storage of nuclear weapons, but made no attempt to limit transit rights on the high seas or port visits in territorial waters. It set out a less demanding treaty than proposed by the coalition of South Pacific anti-nuclear groups — one that it thought would be compatible with United States interests. Some Pacific island states criticised the treaty for being too timid, but in the end, despite Australia's efforts to tailor the treaty to meet United States objections, the USA refused to sign the protocols. The return of cold war tensions in the early 1980s, and the rise of peace and disarmament movements, had affected United States perceptions. The United States administration saw the treaty as part of what it described as a 'nuclear allergy' infecting many parts the world, and particularly New Zealand. Akin to the 'domino theory', the 'nuclear allergy' would spread from state to state and lead to restrictions on United States military deployments unless it was cured (Mack 1988).

REGIONAL ENMESHMENT

The global structures of the British Empire and the cold war have gone, and have allowed Australian foreign policy's current phase of regional 'enmeshment' to develop more directly from Australia's perception of its interests in the region.

The extensive interconnections between Australia and the region have not sprung up overnight but have developed over considerable periods of time and, indeed, through the period of United States alliance. Economic transactions, particularly with regard to Australia's exports, were generated initially by Japan's growth in the 1950s and 60s. This was followed by the growth of other North-East Asian economies, such as South Korea and Taiwan, and then China itself, and transactions with these states were supplemented by expanded trade with some of the ASEAN states, to the extent that about 60 per cent of Australia's current exports are destined for East Asia. Trade flows have been followed by capital flows, with a 'third wave' of foreign investment in Australia from Asian sources supplementing previous waves of investment from Britain and the USA (David & Wheelwright 1989). A growing number of Australian-based firms have located parts of their activities in Asian states.

Looking beyond the economic data, there has been a dramatic growth in the movement of people between Australia and other states in the region. With the abolition of a racially based immigration policy in the period 1966–72, immigrants from many Asian countries have since settled in Australia, usually having competed with would-be immigrants from other parts of the world for a place within government-set target quotas on the basis of skills and family reunion. Others have entered on humanitarian grounds. About half current immigration is from states in East Asia. Moving the other way, about 40 per cent of Australian-born permanent emigrants each year now relocate in Asia. The movement of people as tourists within the region has also expanded dramatically.

Evans's claim that there has been a 'great turnaround in contemporary Australian history' is based in some fundamental and well-established trends (Evans & Grant 1995). These trends give plausibility to the development of a regional concept of diplomacy, based on the premise that 'our future lies in the region'.

The attempt to give Australian foreign policy a distinctly regional dimension, which did not derive from a wider global commitment to Empire or alliance, was also evident before the 1990s. Dr Herbert Vere Evatt, like many other Australian political leaders before him, fore-shadowed the necessity of regional accommodations that would allow the White, settler-colonial nation-state of Australia to exist in relative harmony with the nation-states around it. The postwar Labor government's support for Indonesian independence was evidence of this. Likewise, the Liberal Foreign Minister, Richard Casey, was far more attuned to the region than his Anglophile prime minister, Menzies. However, only since the mid-to-late 1960s has the force of those perceptions become articulated in policy directives. Although he died before he could fully mobilise the resources and commitments to act on it, Harold Holt perceived the importance of 'good relations' with the Asia-Pacific region during his brief term in office. The Whitlam government's ability to begin the process of establishing good relations in the early 1970s was driven more by a changing international order than any driving commitment to regionalism on the part of either the population or its elected representatives. But the United States normalisation of relations with China allowed Whitlam to pursue the ALP's long-held policy of diplomatic recognition.

In the Asia Pacific, the imminent defeat of the USA and its allies' forces in Vietnam, and its subsequent *rapprochement* with China, recast the geopolitical balance in the region. The USA, specifically through Nixon's Guam Doctrine, reassessed its military commitments in Asia and the Pacific. This did not lead to a total withdrawal from the region, but importantly for Australia, it did represent a significant decline in 'dependence'. In a very real sense, United States foreign policy 'forced' the region upon Australia. Australia would have to adjust quickly to these new circumstances. Its involvement in the Vietnam War could not be concluded simply by the withdrawal of its troops. Although the Americans could leave, Australia remained geographically fixed in the region. The task of changing regional states' perceptions of Australia took on a certain urgency.

But only since the end of the cold war in 1988–89 has the most concerted effort been made to build a diplomatic strategy around

Australia's place in the region. It is difficult to dispute Evans's statement that 'the most active, and probably the most distinctive dimension of Australia's diplomacy in recent years has been regional':

> The most visible manifestation of this has been our initiative under Bob Hawke in launching the Asia Pacific Economic Cooperation [APEC] process in 1989. We have played a leading role since, under Paul Keating especially, in expanding APEC's horizons — to the extent that it has gone well beyond data exchange and policy dialogue, and even beyond the specific cost saving strategies we describe as trade and investment facilitation, to the point where it is now about to play an historic role in advancing trade liberalisation. APEC is now accepted both within the region and around the world as the Asia Pacific region's pre-eminent economic forum, and as an engine for potentially very major economic change. If there had been any lingering doubt about its relevance and utility, this was comprehensively dispelled at the Bogor summit.
>
> The other important context in which a sense of community is growing in the Asia Pacific is security. When I first floated, four years ago, the possibility of the evolution in the Asia Pacific region of a new regional architecture — modelled very loosely on the emerging CSCE [Conference on Security and Cooperation in Europe] ... — to respond to the new security realities of the post-Cold War world, I met with a less than enthusiastic response. But times have changed, and what seemed very radical propositions then have now become almost regional orthodoxy. The most important development was undoubtedly the convening in July this year, in Bangkok, of the eighteen-member ASEAN Regional Forum: a new process of dialogue and cooperation embracing all the region's major security players (including all those countries traditionally hostile towards or nervous about each other), and with an agenda that already includes trust and confidence building measures, the development of preventive diplomacy processes and, possibly, cooperative peace keeping activity.
>
> (Evans 1994b, p. 10)

The Labor Foreign Minister took his argument a step further: 'The meetings in Bogor and Bangkok — consolidating and putting in place respectively, both the economic and the security dimensions of a new regional architecture — have made 1994 a watershed, *marking the transition from theory to something very close to reality, of the idea of an Asian Pacific community*' (emphasis added). He elaborated that, by the idea of community, he meant 'the Chinese literal translation sense of big family ... a community, moreover, in which Australia is unequivo-cally seen not as an outsider or bit player but as an accepted involved and participating partner' (Evans 1994b, pp. 10–11).

The Foreign Minister went on to outline the 'East Asian Hemisphere Community', in which Australia was an 'East Asian Hemisphere Nation', and his office produced a map to illustrate the Hemisphere and Australia's place in it.

EVALUATING AUSTRALIA'S REGIONALISM

Australia's regional policy has been highly praised by Joan Beaumont and Garry Woodard, who saw the ALP's 1993 election win as allowing the government 'to reap the harvest of an unusual number of diplo-matic initiatives which had been three to ten years in the making'. These included 'a full hand of Australian foreign policy successes ... : Asia-Pacific Economic Cooperation (APEC); the successful conclu-sion of the Uruguay Round locking agriculture into the GATT system for the first time; a political settlement in Cambodia; a Chemical Weapons Convention; and a regional consensus on the opening of a security dialogue' (Beaumont & Woodard 1994).

Is there a triumphalism in the regional rhetoric of both the Minister and some commentators? Are the praise and self-congratula-tion justified by the achievements? The answers to these questions depend to a considerable degree on the extent to which the achieve-ments have effectively addressed the difficult issues that shape the region's interstate politics. Each element of the regional 'architecture' can be seen as relating to and deriving from a different regional 'sys-tem'. It can be analysed according to its probable consequences for that system and its interaction with other systems.

The ASEAN Regional Forum (ARF) is an element that attempts to address the problems and possibilities of the region as a *threat system*: the problems of conflict and rivalry arising from national sovereignty, national security, and national defence organisations. The forum aims to constrain and to prevent conflict through ideas of 'confidence-building' in the military area, awareness of 'common security', and the like. As an attempt to build order and strengthen the 'society of states', it is essentially an initiative inspired by rationalism.

APEC attempts to address the problems and possibilities of the region as an *exchange or economic system*, in which states can attempt to secure narrow national advantage through economic protection and through win–lose policies towards other players. APEC aims to counter this impulse, and to promote lower protection and win–win outcomes. It is informed by a neo-liberal approach to International Political Economy.

The idea that Australia should see itself as an East Asian Hemisphere nation attempts to invest proximity with meaning and to find an identity for Australia within the region as part of a 'community' or cultural system.

The traditional realist test for both ARF and APEC is whether they can assist in managing major conflicts between the member states of these forums. These conflicts include the dispute among many East Asian states over the Spratly Islands (a dispute that has begun to engage elements of the regional threat system) and the trade war between Japan and the USA (a dispute arising in the regional economic system) (Mack & Ravenhill 1994).

The regional threat system

With regard to military threat, the agenda of ARF is focused on general matters, such as confidence-building and transparency of military developments, which provide the 'building blocks' of common security (Ball 1991). Cooperation may lead to new habits in thinking about interstate conflict. Kenneth Boulding (1978) distinguishes between 'stable peace' and 'unstable peace': in the former, wars between states become unthinkable, in the latter, peace may still be interrupted by periods of armed conflict. If the 'building blocks' do not create the capacity to tackle the thorny issues of potential interstate conflict, and to assist in shifting the region to stable peace, then ARF may have raised false expectations about its promise and potential. The discussions may be confined to the 'soft' security issues, avoiding issues in which clashes of interest are more evident.

A still more difficult test may be managing or resolving *intrastate* conflicts, particularly those with the potential to spill over into interstate conflict. The need to address conflicts within states may simply not be met by a meeting of states who would regard this as interference in 'domestic' matters.

Australia has celebrated the active role it has played in establishing this forum and, at last, being an 'inside player' with regard to military security in the region. Australia's regional security activism contrasts with its response to the Gulf War in 1990, when it speedily agreed to participate without sounding out the views of governments in the region (Malik 1992). There was no exploration of any possible joint initiative between Australia, with its historic ties to the USA, and Indonesia, with its connections to the Islamic world. As in the 1950s and 1960s, the region was ignored when the United States alliance called its dues. With the end of the cold war, some continuing United States military presence in the East and South-East Asian region is seen in positive terms by many states. It is seen as a deterrent to possible attempts by other powers to establish regional hegemony. This leads to a happy concordance of views between regional states and the USA, and as long as this continues, Australia will not face difficult dilemmas. But if and when the security perceptions of the USA and of ASEAN begin to diverge, it remains to be seen whether Australia will continue to develop a regionally based conception of its interests or fall in yet again behind the USA. Australia's regional military security is discussed more fully in Chapter 7.

According to the neo-liberal economic assumptions of Australia's 'regional architecture', the APEC process will reinforce the military security process. Economic progress defuses the potency of the threat system, and cooperation in military security is expected to follow mutual economic gain. Economic links can certainly develop and produce these outcomes. But economic growth also allows states to acquire more sophisticated and expensive military capabilities. Growing military capabilities may undermine progress in common security by creating suspicions among states about the others' intentions. As the lethality of national defence organisations within the threat system rises, the costs of any future breakdowns in common security increase. This is the basic rationale for developing arms control as an element of common security. The continuing pace of military modernisation has not yet, however, been put on the agenda for regional security dialogue, and Australia has not taken any initiative here. Some Australian statements have maintained that Australia's military modernisation gives it a 'technological edge' over other states in the region — hardly a situation that will endure. One consequence of the commitment to free markets is the tendency to regard military acquisitions as simply another form of economic development and trade. The arms trade is losing its morally problematic status and becoming just another form of business.

The regional economic system

APEC is sometimes described as a 'three-layered wedding cake', comprising:

1 data compilation and sectoral dialogue
2 trade and investment facilitation, and
3 most recently, a commitment to trade liberalisation (Evans 1995).

The degree of commitment by APEC members to trade liberalisation is challenged by sceptics, who see nationalist economic agendas asserting themselves strongly. APEC is also challenged by the spectre of an alternative regional organisation, the proposed East Asian Economic Caucus (EAEC), which has been championed by Malaysia, although unsuccessfully to date. Australia has not been included in Malaysian Prime Minister Dr Mahatir's proposals for the group. Malaysia also vetoed an invitation by ASEAN for Australia to participate as part of the Asian group in Asian–European talks in 1996. The EAEC proposal's lack of success has been largely the result of Japan's refusal to join a regional economic organisation of which Australia is not a member. If APEC fails, there is a competitor in the wings. A crucial test for the APEC process is whether it provides a sufficient basis of common interest to moderate substantial economic conflicts between two of its members — the USA and Japan, without Japan making bilateral concessions to the USA at the cost of other states in APEC. The wedding party may not turn up, despite the multi-layered cake.

Without waiting for APEC to succeed in organising multilateral trade liberalisation, the Australian government committed itself to following neo-liberal economic policies at home. It did this in order to capture a share of the benefits of the regional economic sub-system into which it was increasingly drawn. This policy shift has had profound effects on Australian economic structures and society, and its efficacy is discussed in Chapter 8.

Region as community

Perhaps the most adventurous, and the most problematic, idea about the region is that recent developments in military security and economic cooperation have brought the idea of an 'Asian Pacific community ... very close to reality'. Gareth Evans identified additional underlying factors that have led to a high degree of cultural convergence among the countries of East Asia, including Australia: trends in technology, the widespread use of English, and the strength of the concept of liberal democracy (1995).

The idea of community is implicit in most uses of the phrase 'in *our* region'. This phrase seeks to invest proximity or closeness with meaning, based on an idea of shared interests, shared values, and a shared future. To speak of others being 'in our region' is not just a reference to matters of geography — to space and distance — but also a translation of space into place, a statement about who we are as Australians and the identity of those with whom we have, or hope to have, some sort of special relationship.

This idea is problematic for two reasons. First, it cannot define the boundary of the regional 'community' because it cannot settle on a definition of who's 'in' and who's 'out'. Senator Evans's statement entitled 'Australia in East Asia and the Asia Pacific: Beyond the Looking Glass' (April 1995) attempts to define Australia's region both as that encompassed by APEC membership (described as the 'Asia-Pacific'), which includes the USA across the Pacific Ocean, and as the somewhat smaller grouping encompassed by East Asia, which excludes the USA. The USA is initially defined as being 'in' Australia's region: 'the Asia-Pacific … and the US in particular as a crucial continuing economic and security player — would always remain for us, as it should for every country in the region, the main game'. A few sentences later the USA is defined as 'out': 'We looked in the mirror in the early 1980s and began to see us as others had long seen us: politically and militarily dependent on others half a world away'. The USA shifts from being a crucial regional player to being 'half a world away'.

Second, the idea of regional community (whether 'Asia Pacific' with the USA, or the 'East Asian Hemisphere' without it) is problematic because it seems to refer principally to state-to-state relations, as epitomised in the APEC and ARF meetings of heads of state, foreign ministers, and state officials. This is a 'thin' concept of community in comparison to one defined on substantial 'people-to-people' relations. Defining the community in terms of state-to state relations may result in an unwillingness to examine 'regional' issues in which governments are in conflict with significant sections of their populations, and in which there is a total or partial absence of democratic processes that can put popular concerns on the agendas of rulers. International economic forces — those facilitated by the APEC process — are relocating and redefining benefits and risks, including environmental risks, across the region. States may be relatively disinterested in the fate of adversely affected groups, and may actively suppress dissent. Defining regional community in state-to-state terms may allow Australian governments to adopt the same lack of interest in human rights issues as more authoritarian members of the 'community'.

Globalism and neighbourhood

The factors that Evans cites as leading to cultural convergence in the region may have just as much relevance to the world as a whole. The underlying processes of economically driven globalisation would appear to counteract the tendency for particular regions of states to become distinct from others. To what extent is Australia's proximity to East Asia relatively unconnected to its substantial trade focus there? Did trade develop because of economic complementarity, in which closeness was fortuitous but largely irrelevant? Alternatively, did distance-connected factors, such as freight costs, play a more important role in the past than in recent years, as freight is a much smaller factor in value-added exports and in the export of services?

This observation is not to downplay the significance of Australia's economic reorientation of trade towards East Asia, but to point out the opportunistic character of economic relations and the tension between any idea of regionalism and globalism. The idea of an 'East Asian Hemisphere community' may be too small a concept in light of globalising trends.

Paradoxically, it may also be too large a concept, compared with a more traditional concept of regional community as *neighbourhood*. Defining Australia as an East Asian Hemisphere nation appears to cut out neighbours in the island Pacific, and may marginalise the place of New Zealand — areas traditionally referred to, with Australia, as 'Oceania'. Can it be argued, contrary to the pervasive idea that Australia's region extends to Japan and China, that there is potentially more meaningful cultural content in a smaller region — one that only encompasses nearby states, including Papua New Guinea, Indonesia, and other parts of South-East Asia? The argument could not be made on an economic basis, as Australia's trade with North-East Asia eclipses that with South-East Asia, and the Pacific islands have lost their former economic significance. Perhaps this case could be made partly on broad security grounds: that problems in the immediate region will have the greatest bearing on Australia's sense of military and environmental security.

CONCLUSION

This chapter has sought to underscore the major redirection of Australian foreign policy towards a focus on the region, after starting unpromisingly by focusing on concepts of Empire and alliance. Regional diplomacy is based on substantial regional economic connections and regional strategic perceptions. These factors have propelled an extraordinary activism on the part of Australia since the end of the cold war. Australia's regional diplomacy has contributed substantially to the creation of APEC and to ARF, although Australia has not yet been put to the test on issues in which a regional consensus on a particular issue is in conflict with United States policy. The main question we must ask is whether the hype has exceeded the realities, and in particular, whether ideas of regional community are presented in coherent, substantial, and sufficiently well-defined form.

5   Australia as a Middle Power

INTRODUCTION

The shift in focus from 'alliance' to 'region' illustrates part of the dramatic change in Australian foreign policy that has taken place since the end of the cold war. Yet the idea of 'Australia's region' is ambiguous, being redefined to accommodate the foreign-policy agendas — military, economic, and cultural — that are seen to be the most important at any particular time. The more general shift in foreign policy can be described as a shift from alliance to *multilateralism*. Australia's great power alliance relationships have, in most respects, been *bilateral* relationships between Australia and Britain, or Australia and the USA. 'Multilateralism' refers to diplomacy based on multiple relationships and coalitions, and the development and utilisation of international organisations. It is not necessarily connected to events in Australia's region or regions.

MIDDLE POWERS AND MULTILATERALISM

Multilateralism has been a major characteristic of Australian foreign policy at times of great power realignment and alliance uncertainty; it is not just a feature of the end of the cold war. The other notable periods of multilateralism were the end of the Second World War, with the diplomacy of Evatt, and the cold war *détente* that coincided with the Whitlam years.

Australia's multilateralism is linked to the idea that Australia is a middle power. In discussing the issues to which Australia's foreign policy will have to respond to into the twenty-first century, Gareth Evans and Bruce Grant argue that Australia already possesses a strong asset in 'our status as a middle power, with the *capacity that implies for effective action and influence*' (1995, p. 343, emphasis added). The former Australian foreign minister deploys the term 'middle power' frequently and points to examples in which previous Australian governments have used the concept similarly. What *is* a middle power, and what *are* the capacities that it brings to international relations? Is the term 'middle power' quantifiable in any objective and meaningful way?

Intuitively the idea of a 'middle power' makes some sense. Australia is clearly not a great power, nor is it an insignificant power. For Evans, determining a middle-power status is a 'matter of balancing out GDP and population size, and perhaps military capacity and physical size as well, then having regard to the perceptions of others' (Evans & Grant 1995, p. 344). Evans and Grant draw on the work of Carsten Holbraad (1984). Holbraad's formulation, like other attempts to rank states according to particular groupings, embraced formalistic economic measurement — using GDP as the determining yardstick in calculating middle-power status.

Measured in terms of GDP, population, military capacity, or other such criteria, the concept suffers from the all-too-obvious problem of oversimplification. The criticism here is that it is not necessarily the level of GDP but the *composition* of the GDP that is the important consideration. Two states with similar levels of GDP will pursue very different foreign, economic, and political policies if one is dependent on agricultural production for revenue and the other on the export of manufactures. Different states have different patterns of historical, ecological, demographic, cultural, economic, political, and social development. And it is these elements, which underpin the GDP equation, that are important factors in determining a nation-state's capacity and willingness to act on particular foreign-policy issues.

A more systematic definition of middle power is offered by Cooper, Higgott, and Nossal. These scholars locate their analysis of two 'middle powers' — Canada and Australia — within the context of a changed international political and economic environment. They observe that since the mid-1980s the 'more traditional foreign policy concerns of a military-strategic nature were increasingly replaced by a mounting concern over the future of the international economic system' (Cooper et al. 1993, p. 4). Eschewing 'traditional definitions' of middle-power behaviour — size, population, power, or geographical location — Cooper et al. define their notion of middle power as 'based on the technical and entrepreneurial capacities of states' that provide 'initiative-oriented sources of leadership' (1993, p. 7). Kim Nossal (1993) takes up this 'Initiative' aspect as one element among five that could be deployed to enhance the task of defining what characteristics a middle power may possess. Nossal argues that of

> special importance to the activist style of middle power statecraft is the diplomatic initiative (usually with a capital 'I'). Typically, the Initiative will involve the middle power making a concerted effort to think through an international problem; generating a plan of action, often based on technical expertise; gathering support for its ideas from as many like-minded states as possible; and then presenting the great powers with a suggested set of solutions, or with a process that might lead to a political solution.
>
> (1993, p. 214)

This type of 'activist style', Nossal claims, propels the middle power towards operating within multilateral forums, which 'provide a legitimate entrée for smaller states into the affairs of the international community as a whole, a voice that would otherwise be denied them' (1993, p. 215).

Robert Cox extends this focus on the middle power's predilection for operating within a multilateralist framework:

> In modern times, the middle-power role ... has become linked to the development of international organization. International organization is a process, not a finality, and international law is one of its most important products. The middle power's interest is to support this process, whether in the context of a hegemonic order or (even more vitally) in the absence of hegemony. Commitment to the building of a more orderly world system is quite different from seeking to impose an ideologically preconceived vision of the ideal world order.

Reinforcing the point about the middle-power need for *order*, Cox claims that such states exhibit 'a commitment to orderliness and security in interstate relations and to the facilitation of orderly change in the world system [and that these] are the critical elements for the fulfilment of the middle-power role' (Cox 1989, pp. 826–7).

What distinguishes Cox's conceptualisation from other discussions of the middle-power thesis is his recognition that the interests of the middle powers are tightly bound to international order. It is, therefore, in these states' interests to promote normative interstate behaviour through the framework of international law that seeks to facilitate cooperation and conflict resolution — the rationship agenda. As Cox notes, this behaviour is likely to increase during times of transition — for example, during the breakdown of hegemony. This important insight goes some way in explaining why states like Australia involve themselves in the multilateral processes.

MULTILATERALISM: CONTEXT AND SETTING

Australia's position in the international system is on the fringe rather than at its core. Australia does have the capacity to influence some external events, but only in particular contexts. Generally Australia has little unilateral influence at the international level. Australia is also particularly vulnerable to sudden changes in the international system.

In considering the ability of Australia to shape the structure of the interstate system, the realist perspective — with its focus on pursuing interests through military power — is perhaps less relevant than the rationalist perspective, which seeks order by way of institutionalised conflict-resolution and negotiation. From the rationalist perspective, states who lack the power to exercise unilateral influence can still play a multilateral role. But what are the avenues that are available for it to protect and further its interests within a specific context of international relations?

In those periods when a strong alliance structure has existed, Australia has often pursued a decidedly realist foreign policy. Sometimes this ultra-realist position has placed Australia at the outer extreme of its principal ally's foreign policy, making Australian foreign policy even more conservative than that of the USA (Leaver 1990, p. 22). Australia's political leaders have sometimes asserted that they have acted as a brake on their principal ally during episodes of escalating confrontation. Where alliances either have not existed or have been poorly defined, Australia's foreign policy has tended to emphasise a strong moderating role by fostering multilateral solutions to issues of conflict. On these occasions it is possible to argue that Australia's foreign policy has been informed by the rationalist perspective. The particular global political context, issue, and level of perceived interest involved have determined whether Australia's foreign policy has been guided by realist or rationalist assumptions about the foreign-policy options that are most appropriate for the enhancement and protection Australia's interests (Indyk 1985).

During periods when a dominant state has been able to establish global control through military and ideological coercion — periods of hegemonic order — 'acceptable' rules and patterns of interstate behaviour have been defined, articulated, and often brutally enforced. For example, taking the period from the postwar years through to the early 1970s as the generally accepted period of United States hegemony, it is clear that the international political economy, international political institutions, and Western military–security alliance networks were developed to support the interests of the USA (Ruggie 1994).

Within that political, economic, and military framework, Australia was able to operate its foreign policy with some assurance that, if attacked, assistance could be sought from the alliance partner. Whether a suitable response would be forthcoming was never certain, and foreign policy was preoccupied with the payment of premiums on the alliance 'insurance policy'. Since the early 1970s there has been a qualitative and quantitative diminution of United States hegemony. In response to these changes, Australian policy-makers have sought, wherever and whenever possible, the promotion of a predictable and orderly international political system.

In an effort to halt, slow, or manage its decline, the waning hegemonic state, with its diminished ability to exercise military or economic leverage, may attempt to reinvigorate the rule- and norm-generating international political institutions and structures that were created at the height of its power. To secure its interests with diminished resources, the declining power may attempt to incorporate other states into 'power-sharing' arrangements that it would not have countenanced previously. Such 'power-sharing', as Richard Leaver (1993) has noted, can often turn out to be little more than 'burden-sharing'. Sharing the former hegemon's burden places allied states in a difficult position.

The disintegration of the 'old' order effectively opens up the multi-lateral processes and allows lesser powers an opportunity to play a role in managing the international system. The costs of being involved may, however, far exceed the potential benefits. Commitments may not match expectations, and particularly where interests do not over-lap, there is likely to be strong resistance to becoming involved in situations thought to be peripheral to the core interests of either state. Indeed, former allies may find themselves at loggerheads on particular issues as the diminution of power and resources constricts the available room to manoeuvre. Relations between former allies thus becomes additionally complicated during periods of hegemonic decline.

Against this backdrop, Australian policy-makers have been presented with a stark choice. Three policy options are available. They can adopt an isolationist foreign policy, knowing that, if there is the threat of conflict, there will be little chance of assistance from their former ally. The second option is to remain loyal to the 'bitter end'. While this option has some rhetorical value in the initial phase of transition, it tends to lose its saliency over the longer term as interests diverge more acutely and resources become targeted more narrowly. The third policy option is to assist in managing the transition period by facilitating cooperation within multilateral forums. In the post-cold war political environment of the 1990s, where there is a perception that the UN has been 'liberated' from bipolar rigidities, this latter option has become the focal point of Australia's foreign relations (Chater 1995, p. 157).

Unlike the relative stability and consistency of the cold war years, this current period of international relations has created a variety of issues to which Australian policy-makers must now respond. Managing foreign policy under these circumstances requires (perhaps demands) a deft hand on the tiller. It is within this wider context that Australia's current foreign policy can be best understood. With the international system now in a period of transition, Australia can, arguably, do little other than respond to the political changes occurring at the global level. This structural requirement to act is nevertheless influenced by policy-makers' decisions. Which particular options are pursued, how the state formulates its foreign policy, and what perceptions and theoretical perspectives inform the selection of particular policy choices over others are based on value judgements made by policy-makers.

The Australian Foreign Minister from 1988 to 1996, Gareth Evans, sought to develop his own particular brand of foreign policy. When first articulating what was to become an important feature of Australia's foreign policy under his guidance — the principle of 'good international citizenship' (GIC) — he was quick to make the qualification: 'Good international citizenship is perhaps best described, not least for the cynical, as an exercise in enlightened self-interest: an expression of idealistic pragmatism' (Evans 1989b, p. 13). But how much does the GIC idea differ from Gough Whitlam's 1973 comment that Australia needed to play an 'enlightened role in world affairs'? (House of Representatives, *Debates*, 24 May 1973, p. 2645).

Similarly, there are clear resonances between Evans's GIC and Bill Hayden's location of Australian foreign policy within 'a global moral consensus' (House of Representatives, *Debates*, 26 November 1985, p. 3665). How should we interpret such statements? Are they positive reflections of a developing maturity and independence in Australia's foreign policy approach since the early 1970s (Bell 1988)? Or are they just the recent manifestations of the politics of threat and fear (Pettman 1992)?

HISTORICAL PATTERNS: THREE PERIODS OF TRANSITION

One method of conceptualising history, in this case the historical development of Australian foreign policy, is to sort significant parts of that history into periods by marking out events that appear to separate or divide. The difficulty is knowing precisely where to draw the line of demarcation. After all, history does not come neatly packaged into digestible sections; history is fluid. Yet there are periods that, in relatively compact time frames, force the revision and reformulation of foreign policy, often as the result of structural change. These are periods of transition. Identifying such episodic nodes is not too difficult. In some cases, these transitions occur when the state is confronted with the failure of its foreign policy — a time of intense crisis. The threat of military invasion, and with it the loss of territorial and political sovereignty, is an extreme example.

In Australia's recent history, there has only been one such case. Prime Minister John Curtin's December 1941 address, in which he appealed to the USA for military assistance 'without fear or favour', has often be seen as the point at which Australia's foreign policy shifted from its reliance on Britain as Australia's 'protector' to dependence on its new Pacific ally. Curtin's address needs to be viewed in the context of Britain's wartime military commitments and constraints. The USA was the only power with the resources to push back the Japanese military. It also needs to be remembered that the USA entered the war because of the Japanese bombing of Pearl Harbor. In that sense, the war in the Pacific was predominantly the USA's concern, and United States military strategy required secure bases and supplies. Australia provided these. United States assistance to Australia during the war was, therefore, bound up in mutual or overlapping interests.

Although times of intense crisis do trigger significant reassessment, these periods are, in a sense, artificial, with the immediate threat taking on an all-consuming importance. It is only when the crisis has ebbed and the intensity has diminished that the policy-makers can turn their minds to reconstructing a foreign policy that will protect against a repetition of the past events. Thus it is not at the times of immediate and intense crisis that the student of foreign policy can find patterns or themes that reveal ongoing constraints and opportunities to reformulate foreign policy. Crisis may well generate change, but there is a lag between the time of crisis and the reformulation of policy.

The important phases occur when Australia's policy-makers are forced to consider foreign-policy options in situations in which Australia and its primary ally have no overlapping interests. This does not entirely invalidate notions such as 'protector', or a 'great and powerful friend' or 'patron'. In the world of states, interests are always dynamic and often highly divergent, even between states who are able to genuinely profess a special relationship. There is nothing in the state system that can guarantee protection from aggression. Friendly states may condemn an attack, but unless their interests are threatened, it is unlikely that they will intervene to support the victim.

Bearing this discussion in mind, three periods of transition in the postwar system can be discerned. Again it is important here to note the lack of a clear definition. The first phase can be identified as a five-year period in which several important events occurred 1945–50. The second phase began in the late 1960s and early 1970s and lasted through to the mid-1970s. The final period is closely aligned with the end of the cold war: from 1990 to the present. To underscore the point once more, these are periods in which Australian policy-makers found themselves having to respond to major changes occurring at the system level.

TRANSITION 1: FROM EVATT TO ANZUS

In the period between the end of the Second World War and the early 1950s, Australia's political leaders were confronted by the immediate problem of how to formulate a foreign policy that would offer the highest level of security. Sandwiched between the wartime military commitment and the tenuous, but nonetheless real, ratification of the ANZUS treaty, this half-decade contained some important indicators of the nature of Australian foreign policy.

One possible reading of this period is that Australia's political leaders simply recognised the reality of the moment. British power had all but collapsed, and thus the logical position was to find another 'protector'. Moving from the scant protection offered by the defoliated 'oak' to the shelter of the sturdy 'redwood' is a captivating but inadequate metaphor for this phase. The transition was not that simple and did not fully take root until the 1960s. Although the USA held the atomic monopoly, it was only a matter of time before others would obtain a similar capacity. Atomic arms racing was a strong possibility. And as was the case at the end of the First World War, by the close of the Second World War many felt that the system of power politics and interlocking alliances that had been such a prominent feature, if not the catalyst, of both wars should not be allowed to reassert itself.

In the immediate postwar period, Australia's political leaders believed that participation at the UN would enhance their ability to secure what they perceived to be Australia's interests. This participation was never promoted as an alternative to engaging in alliance politics with the USA, Britain, or both. The image that emerges, therefore, is one in which Australia's political leaders were using all possible levers in an attempt to secure perceived interests and operating to insure against any unexpected contingency that might arise. At times, as Carl Bridge notes, this process unleashed contradictions, but these were never allowed to undermine the fundamental importance of the overarching alliance structure (1991, p. 6).

Having faithfully served the allied cause during the Second World War, the ALP's wartime foreign minister, Dr Herbert Vere Evatt, believed that Australia had earned a special place in the postwar order. 'In the Pacific', Evatt had argued, 'we fight not for ourselves alone but as trustees for the United Nations, particularly for the British Commonwealth of Nations' (Evatt 1945, p. 130). As Evatt readily acknowledged, Australia was 'vitally concerned in the establishment of a successful peace and world security system' because of its 'vulnerable position' (Evatt 1945, p. 210).

Furthermore, stressing the important function that states such as Australia could perform within this new order, Evatt argued, 'It must be remembered that a so-called small Power may in certain areas and in special circumstances possess great, if not decisive influence' (Evatt 1945, p. 212). It was hardly surprising, therefore, that the Australian–New

Zealand ANZAC Agreement (Canberra Pact) was situated within this broader context. Evatt argued:

> It would be wrong to contend that Australia and New Zealand can have an exclusive concern with the future of any part of the Pacific region. In particular, without the continued interest and active participation of the United States (as well as the United Kingdom) in arrangements for welfare and security, there is no hope of stability and harmonious development in this area.
>
> (House of Representatives, *Debates*, 30 November 1944, p. 2536)

In guiding Australia's foreign policy, Evatt chose a course that he perceived would enhance Australia's security, not through a simple alliance agreement with any particular 'great power' — Britain's inability and the USA's unwillingness left few options — but within a layered system. This policy approach found expression in both Evatt's support of multilateralism and in his desire for a strong regional alliance. Although Evatt's multilateralism was focused on the UN, this was not to the exclusion of the British Commonwealth. Nor was it the case, in Evatt's mind, that multilateralism and alliances were mutually exclusive policies.

Evatt's assessment was developed as a way of protecting what he perceived to be Australian interests in the postwar global context. His discussion of United Nations Trusteeship — Australian political control of Papua New Guinea — was as much about 'forward defence' and the 'strategy of denial' as it was about transmitting positive messages to a wider audience that Australia was willing to assist in the decolonisation process and to accept its responsibilities as a regional power. Similarly, his successful promotion of the domestic jurisdiction principle in the UN Charter was bound up in a recognition that it would not be in Australia's interests to have its human rights record discussed openly within the UN, particularly when Australia was attempting to establish durable relations with neighbouring states for which the issue of race had a specific poignancy.

Evatt's decision to reformulate Australia's foreign policy in the five years after the end of the Second World War was rapidly overtaken by events beyond Australia's control. The cold war became entrenched, and East–West tensions shaped the 1950s and 1960s. Australia's foreign-policy-makers did not withdraw totally from multilateralism to alliance politics as ANZUS stamped its mark on foreign policy. But gone was the earlier sense of urgency in using these multilateral forums as rule-governing institutions of interstate behaviour. Expectations generated by the all-too-brief multilateral moment of the mid- to late 1940s soon dissipated as the UN Security Council became yet another arena of superpower rivalry.

TRANSITION 2: FROM WHITLAM TO FRASER

In the period between the late 1960s and the early 1970s, the international system again underwent fundamental change. Beneath the political *détente* in United States–Soviet relations and a *rapprochement* between the USA and China, there was a growing awareness that the postwar international system was beginning to crumble. In addition, rising global inflation, a by-product of the war in Vietnam (Riddell 1989), further undermined confidence in the economic power of the USA; this economic pessimism was accompanied by the all-too-obvious looming military defeat in Vietnam.

President Nixon's Guam Doctrine, announced in July 1969, aimed to substantially reduce the USA's military commitment in the Asia-Pacific region — Korea excepted. The USA's unilateral action ending the Bretton Woods international monetary system in 1971 — the foundation of the international economy since the late 1940s — pointed to the deepening crisis in the United States economy. United States hegemony was clearly declining (Mack 1986; Foster 1985; Strange 1988). For a nation like Australia, which had been inextricably woven into the fabric of *pax Americana*, the new realities were deeply disturbing.

Coupled with the British 'East of Suez' policy some two years before, the Guam Doctrine articulated what was essentially a 'West of Hawaii' policy. This signalled a reassessment of 'great power' military involvement in Australia's region (Harper 1976). Since this region had always been of primary strategic importance to Australia, the reduced military commitment of the USA and Britain required Australia to reconfigure its relations with the surrounding states. In contrast to the principally strategic interest of the 1950s and 1960s, the relationship between the region and Australia was now moving towards one of greater political and economic involvement (Camilleri 1973). This was the new political environment in which the Whitlam government would have to refashion Australia's foreign policy (Mediansky 1972; Miller 1974a; Miller 1974b).

Multilateralism offered the Whitlam government an opportunity to adjust its foreign policy to accommodate the new regional and global realities without moving too far away from the established and familiar political institutions, which still reflected the values and politics of the 'old order' (Albinski 1977, pp. 248–51). Whitlam's success in redirecting Australia's foreign policy was, however, largely a matter of timing. As he later acknowledged, 'had we come to power in 1969, all our initiatives in foreign policy would have been much more difficult to achieve and much more violently opposed than they were' (Whitlam 1985, p. 26). The changes wrought by Whitlam were based on a *realpolitik* assessment of the changing international order (Bull 1975a; 1975b). Regional issues were important within that process.

The acknowledgment that Australia's future was increasingly bound to the region required not only the formulation of sound foreign policy, but also a serious reassessment of Australia's domestic policy (Hastings 1977). Rather than moving to initiate significant change, Whitlam adopted a middle-course strategy. He sought to maintain existing security and economic ties with Australia's traditional partners, while at the same time cautiously building diplomatic and economic links in the region. In order to accomplish this, Whitlam sought to recast Australia's image. An attack upon racism was an important part of this strategy.

Acknowledging a racist history and speaking out against racism as practised in other states became an important, if not a fundamental, part of the Whitlam government's foreign policy (Altman 1973, pp. 101, 105–6; Lonie 1971, p. 66). His argument that support for racism abroad was tantamount to acceptance of racism domestically galvanised Australian foreign and domestic policy (*Sydney Morning Herald*, 14 November 1972). Condemning racism in southern Africa offered the most appropriate vehicle for Australia's new diplomacy. It would be too simplistic to suggest that Whitlam targeted southern Africa because it was an area of diminishing importance for Australia and one that was far removed from Australia's immediate region. At the time, the White, minority-ruled state of Rhodesia was the subject of UN Security Council sanctions (Polakas 1980). Supporting multilateral sanctions against Rhodesia allowed the Whitlam government to express its abhorrence of racism, as well as to be seen to fulfil its international obligations (Goldsworthy 1975, pp. 1–5; Clark 1973; Clark 1974), simultaneously legitimating and reinforcing the processes of multilateralism — processes that were becoming increasingly important in the international system.

Significantly, Australia shared a common heritage with Rhodesia and South Africa: all three were White settler colonial states with a similar imperial connection. By openly condemning the political structure of these White, minority-ruled states, Whitlam was able not only to establish credibility by attacking 'kith and kin', but also clearly placed these states beyond the pale of international standards, so overcoming what John Miller once termed the problem of 'choice and association' (1971, p. 134). Local racism, whether in Australia or in the region, was of a lesser degree than the particular variant of racism to be found in southern Africa.

Promoting an anti-racist policy in this manner, Whitlam sought to realign Australia's foreign policy with the new realities of a changing international environment (Willesee 1975, p. 7). Here domestic and foreign policies presented themselves as two sides of the same coin. Whitlam's apparent success stemmed not from the abandonment of traditional methods of statecraft, but rather from a clear and accurate appraisal of what was perceived to be in Australia's national interests and how best to achieve these goals (Camilleri 1973, pp. 11–13). Against the backdrop of Australia's muted support for White, minority-ruled states during the 1960s (Menzies 1967, p. 192; Howson 1984 p. 176), Whitlam's redirection of policy did appear to be a 'leap' rather than the 'twist' that it really was (Mackie 1976).

When Malcolm Fraser was elected prime minister in 1975, he continued with many of his predecessor's foreign policies. In some areas, the Fraser Liberal-led coalition government strengthened and extended the foreign policy of the Whitlam era. In particular, Fraser played an important role in the negotiations that eventually led to the creation of the independent African state of Zimbabwe (Neuhaus 1988; Ingram 1979, pp. 275–83). The Fraser government appeared to view the Commonwealth as the most appropriate multilateral institution to further Australia's interests (Higgott 1981, p. 228). Perhaps, like the ALP's wartime prime minister, John Curtin, Fraser also thought that 'Australia was a bigger fish in the Commonwealth pond than it could ever aspire to be in either the American or the United Nations ponds' (Day 1991, p. 68). Perhaps, too, Fraser was less concerned that the Commonwealth was, as Martin Wight once asserted, 'a ruminant, not a carnivore, in the international jungle' (1978, p. 123).

The area in which the Fraser government did diverge sharply from the foreign policies it had inherited was in relation to the USSR. Fraser's preoccupation with a perceived Soviet naval build-up in the Indian Ocean generated significant difficulties with a number of regional states. Fraser's warnings to United States President Jimmy Carter about the need for the USA to reassert its regional dominance not only offended many of Australia's neighbours, but it also failed to elicit the sort of response Fraser had originally sought (Girling 1977; Cheeseman 1993, pp. 9–12).

TRANSITION 3: END OF THE COLD WAR — INITIATIVE-TAKING AND COALITION-BUILDING

During the Hawke–Keating period, patterns of multilateralism continued and strengthened. Multilateral initiatives under the ALP's foreign ministers in this period, Bill Hayden (March 1983–August 1988) and Gareth Evans (September 1988–February 1996), reached new levels. These political leaders sought ways of legitimating and rationalising their chosen foreign-policy directions. Not surprisingly, the ALP made much of the 'Evatt tradition' (specifically his role as one of the architects and founders of the UN) and used the middle-power terminology with ever-increasing frequency. However, the international system has undergone substantial changes since the early 1980s, when the ALP first came to office.

It was not until the mid-1980s that the Hawke government began to fully engage in multilateral processes. During the Reagan presidency, the USA actively blocked any substantial movement within the UN and in any other international forums in which it carried influence. There was little incentive for the Australian Labor government to continue pushing for policy change at the international level, where the USA considered that its interests would be undermined by any multilateral agreements not tightly bound to its preferred policy outcomes. Nevertheless, the Hawke government persisted with multilateral initiatives. In so doing, the government consciously attempted to pursue its multilateral objectives without alienating its powerful ally.

Bill Hayden sought to articulate this difficult position:

We are part of a discernible community of nations with the flexibility and independence to refuse to be limited by what the super-powers and their conservative followers might insist is the iron logic of a purely bipolar world system ... This is why we involve ourselves in the great international social issues.

This is why we set ourselves the continuing task of building bridges between groups, particularly in the critical area of disarmament. This is why we sought membership of the Security Council and have offered our strong support for multilateral solutions to the urgent issues of our time.

(House of Representatives, *Debates*, 26 November 1985, p. 3666)

Hayden chose his words carefully. His criticism of the 'super-powers' avoids any direct reference to the USA at a time when President Reagan's cold war rhetoric had reached its zenith — to the extent that is was threatening global instability. Moreover, he highlighted the role that Australia could play in these difficult times by 'building bridges' and emphasised Australia's support for 'multilateral solutions'.

Faced with United States intransigence at the UN, the Hawke government pursued its policy within other international and regional political forums — the Commonwealth and the South Pacific Forum. Nonetheless, where these forums articulated policy that was perceived by the Reagan presidency as counter to United States interests, the government moved quickly to avoid possible conflict with the USA. The turning-point came in 1986. When Reagan and Gorbachev met to begin talks aimed at restricting the further production of nuclear weapons and reducing existing stockpiles, the Hawke government embarked on a range of initiatives designed to facilitate the disarmament process.

Casting itself in the role of 'honest broker', the Hawke government quickly set about the process of coalition-building. The government launched a series of international initiatives that it perceived to be important to Australia's interests. Among the list of initiatives that the Labor government claimed to be significant were the Chemical Weapons Convention (CWC), the Madrid Protocol (to the Antarctic Treaty), the Australian peace proposal on Cambodia; and the Cairns Group (agricultural reform of the GATT agreements). Gareth Evans would argue that these are policy initiatives that distinguish a middle power (Evans & Grant 1995, p. 346).

Evans acknowledged that these foreign-policy initiatives were in response to international changes. Moreover, he was keenly aware that the pace of reform is likely to be unrelenting:

We will not be able to linger on past achievements; nor will we have the luxury of stepping back from the flow of events. We need now to build on our achievements, develop the capabilities we have established, and have the stamina to pursue favourable outcomes to the many courses of action already initiated, and the many new activities which the future holds. And we need to do this in a way which is sensitive to the particular currents and nuances of our region.

(Evans & Grant 1995, p. 355)

THE LIMITATIONS OF MULTILATERALISM

Australia's current foreign policy draws upon a strong tradition of multilateralism. This tradition is not necessarily a Labor tradition, but Labor governments have been in office at times when multilateral diplomacy has been encouraged by external events. The Liberal-led coalition has nevertheless articulated a multilateralist approach during its recent period in the wilderness. Multilateralism has been very much a part of the rationalist approach at times when the international system has been in a period of transition. When there has been a higher degree of stability within the system, the tendency has been to revert to alliance politics.

There can be little doubt that the end of the cold war brings opportunities for states such as Australia to take a more global approach in their foreign relations. However, the historical analysis presented here, particularly the focus on the three transition phases, signals that current foreign policy, while rating high on the activism scorecard, is yet to show a significant swing away from traditional methods of statecraft.

An international institution is a power structure, whether it be the UN, the General Agreement on Tariffs and Trade (GATT), or its replacement institution, the World Trade Organisation (WTO). As such, these institutions are not forums based on the equality of their member states. Within these political institutions, key states have the power to shape and determine policy outcomes.

A case in point is the UN. Effective action undertaken by this body is determined by the permanent members of the Security Council. Suffice to say that these states take positions determined principally by their own interests. Therefore, middle-power influence is highly contingent on the interests of the great powers. Where interests between great and middle powers diverge, it is hardly surprising that the interests of the former take precedence over the interests of the latter. Gareth Evans's attempt to generate a program of reform of the UN was set out in his book *Cooperating for Peace* (1993); the program comprised a raft of reforms that he argued would strengthen the institution and make multilateralism work more efficiently (Lawson 1995).

The ability to pursue these initiatives effectively is dependent upon the relative openness of the multilateral process. However, there are dangers for Australia in looking to the UN as the mainstay in its foreign-policy initiatives. While the re-emergence of cold war politics is highly unlikely, there is a strong possibility that the UN will yet again become emasculated by the politics of its major benefactor. As Evans notes, the Republican victory in the 1995 congressional elections in the USA means that 'the bottom has fallen out of the UN reform market'. Has Australia placed too much emphasis on multilateralism and yet again missed the opportunity to establish strong and durable bilateral relations (other than those it has attempted to form with traditional allies)?

CONCLUSION: 'MIDDLE POWER' — AN UNDER-DEVELOPED CONCEPT

A difficulty with the orthodox concept of 'middle power' is that it posits a range of static and particular interests that are supposed to shape the policy-formulation process. Questions about how these interests conflict with the interests of other states, particularly allies, or how they might assume a specific importance in a changing international political environment, are left unexplored. A similar problem exists with the recent literature on multilateralism. Much of this work has been particularly narrow in scope. Overall, most of this scholarship could be loosely categorised as dealing with the multilateralism of United States interests (Keohane & Nye 1985; Keohane 1990; Ruggie 1993; Karns & Mingst 1990). Absent is any strong sense of a structural explanation that can account for the changing nature of interests.

At the descriptive level, the idea of 'middle power', as a measure of capacity, holds little value. Attempts to precisely measure or quantify what defines a middle power soon collapse into subjective claims about a state's level of foreign-policy activism. Similarly, at the prescriptive level, it has been shown that, while states pursue particular interests, which on occasion may overlap with others, there is no determining logic of 'middle-powerness' that will bring these states' interests into conformity. States' interests more often diverge than coincide. Why, then, does the idea of 'middle power' hold sway within policy-making elites?

The appeal and utility of the middle-power concept for policy-makers is its durability. As a general, although poorly defined and specified, descriptive term, 'middle power' offers policy-makers a constant yet suitably mutable explanatory framework to rationalise and legitimate their policy choices. The concept of 'middle power' overlays the realist-versus-rationalist discourse. Conceptually, the middle-power thesis slides easily between alliance and multilateralism.

Where an alliance structure dominates, for instance, the middle power concept can be deployed by political leaders and policy-makers to explain the prevailing international power structure and Australia's role within it. Australia can be a 'middle-power ally'. Here notions of regional policing as part of a greater alliance structure dominate. Yet when the system changes, the image of a middle power is transformed into that of a multilateral actor. While the form is elastic — thus providing flexibility for the policy-makers — the label remains the same, and as such, this maintains the all-important appearance of policy consistency and continuity.

Both realist and rationalist conceptualisations of the middle power offer policy-makers a role to play. Policy-makers need to legitimate their existence, particularly in liberal-democratic states, as much as they need to rationalise their policy perceptions and choices to domestic and international audiences. In that sense, the idea of a middle power is as much about the relationship between the policy-makers and the population as it is about making foreign policy. Policy legitimation is thus an important aspect of foreign policy. In foreign policy there is a strong requirement to establish a high degree of consistency across time and contexts, and between changes in government. The 'middle-power' idea thus becomes a useful explanatory framework that allows policy-makers room to manoeuvre. Devoid of a specific form, and not bound strictly by context, the idea of 'middle power' is used to 'sell' foreign-policy decisions.

Domestic consensus becomes particularly necessary during periods of transition, when the population will be required to bear the costs of foreign economic and political policies that generate significant forms of domestic economic restructuring. Policy-makers are required to cooperate, collaborate, and more importantly, compete with other states. Policy-makers who recognise the limits of their state's unilateral power are reduced to trying to build coalitions of overlapping interests. It is at the level of international policy-making elites that the middle-power thesis finds its real niche. Policy-making elites can use the 'idea' of shared interests as a way of communicating across borders in an effort to build coalitions of like-minded states and to encourage commonality of action.

This political utility comes at the cost of analytical and conceptual utility. The ambiguity of the 'middle-power' idea — between a realist/alliance-oriented idea and a rationalist/international order-oriented idea — may inhibit Australia from developing a more strategic coherence in its multilateral diplomacy. Evans's concept of 'Good International Citizenship' (GIC) seems to reflect these limitations.

On the one hand, the statements about GIC can be seen as part of the rationalist discourse in which cooperation, mediation, and an ability to play a constructive role in international affairs are vital components of the multilateralist agenda. On the other hand, these statements can also be viewed as being little more than legitimating rhetoric for a foreign policy that is still overly reliant on realist perceptions (based on threats to sovereignty) and prescriptions (self-reliance). For example, despite the promotion of human rights as an important feature of today's foreign policy, critics of GIC would argue that these concerns fade when the hard issue of commerce intervenes (Goldsworthy 1994). The fact that Evans lists GIC third in his summary of Australia's national interests seems to indicate that these issues are accorded a relatively low priority, rather than the commitment to a thoroughgoing rationalist, multilateral agenda that the concept appears to imply, and that the post-cold war (dis)order may require.

But Howard romore the conceptallogether?

Part III

Issues in Australia's
Foreign Relations

INTRODUCTION

In the first half of the twentieth century, global military insecurity manifested itself most strikingly in the two world wars. The war of 1914–18 was called the 'Great War' by contemporaries in recognition of its unprecedented geographical scope and scale of destruction. This left the conflict of 1939–45, which was fought in even wider domains across the globe, and with much more advanced technology, without a name. The Great War was renamed the First World War, and the war started and lost by the fascist powers — Germany, Italy, and Japan — became the Second World War.

The brief period of optimism about the new United Nations organisation and 'collective security' had ended within a few years of the end of the Second World War. The cold war conflict between a United States-led Western alliance and a Soviet-led communist bloc introduced a radical new dimension of global insecurity. The new weapons of mass destruction, which had been developed by the USA and used on two Japanese cities in 1945, became the counters in the superpower arms competition that developed between the USA and the USSR.

The weapons of mass destruction challenged the principle of discrimination between military targets and civilian population, between combatants and non-combatants, between munitions factories and primary schools, as demonstrated to all by the destruction of Hiroshima and Nagasaki. The size of test detonations grew from the equivalent of tens of thousands of tonnes of TNT to H Bombs with the explosive force each of a 'thousand Hiroshimas' — a prospective level of devastation that seemed beyond comprehension. The effects of nuclear fallout after the initial blast were then found to be far more extensive than initially thought, further adding to the prospects of indiscriminate killing. Intercontinental ballistic missiles (ICBMs), deployed in the late 1960s, enabled each warhead to dispatch a 'thousand Hiroshimas' from one side of the world to the other, or from submarines. Some prominent scientists speculated that a series of nuclear explosions would lift so much material into the stratosphere that it would block sunlight and lower the earth's temperature, producing a 'nuclear winter', agricultural collapse, and the consequent collapse of societies across the Northern hemisphere or the whole planet (Harwell 1984).

The arms race between the two superpowers was described as 'vertical proliferation': an increase in the number and quality of warheads held by the USA and the USSR. There was also 'horizontal proliferation': the spread of nuclear weapons to an increasing number of states. Two of the USA's allies, Britain and France, developed their own nuclear weapons, and China, allied to the USSR for about a decade after 1949, successfully tested the bomb in 1964.

The cold war did not escalate into the Third World War, a prospect widely regarded as having the potential to destroy most of the planet. The adversaries were apparently deterred by this very prospect.

In the post-cold war world, there has been a reversal in vertical proliferation on the part of the USA and the successor states to the USSR. With the breakup of the USSR, four states were left with nuclear weapons: Kazakhstan, Ukraine, Belorussia, and the Russian Federation. The first three agreed to relinquish all nuclear weapons in return for security guarantees from the USA and Russia. The Russian Federation implemented Strategic Arms Reduction Treaties (START 1 and START 2) with the USA. START 1 reduced strategic arsenals from about 10 500 warheads on each side to 6000. START 2, signed in 1993, sought to reduce each total further to 3500 by the year 2003. United States President Bill Clinton observed that:

> together these treaties will leave the United States and the former Soviet Union with only a third of the warheads they possessed at the height of the Cold War. They will help us to lead the future to a direction we have all dreamed of, one in which the nuclear threat that has hung overhead for almost half a century now is dramatically reduced.

> (*Nuclear Proliferation News* 1995, issue 17)

These improvements in the climate of global *military* security are often contrasted with new *non-military* agendas of global insecurity:

- global environmental issues such as atmospheric warming and ozone depletion
- economic globalisation, in which the distribution of risks and benefits is a function of the 'division of labour' and of the distribution of capital in the global economic system, and in which large parts of the planet live in poverty
- human rights violations on a large scale.

These issues and their impact on Australia are discussed in subsequent chapters.

Unfortunately, global military security is still on the agenda, despite the nuclear arms reductions and the rise of other pressing global issues. Two dimensions of global military insecurity in the post-cold war world arise from weapons of mass destruction, and violent interstate and intrastate conflicts.

ARMAMENTS

One set of issues revolves, still, around weapons of mass destruction, and in particular around the problems of continuing horizontal nuclear proliferation and inadequate vertical de-proliferation. Under START 1 and 2, the USA and Russia are reducing nuclear warheads to a cap of 3500 each. But does this represent a fundamental improvement in global security as President Clinton and others have maintained? There still appears to be enough explosive power to turn much of the world into rubble and then make the rubble bounce. Some strategic analysts argue that the objective should be to reduce the United States nuclear capacity to the level of the 'minimum deterrent', but is this achieved at 3500, 350, or thirty-five warheads? Since the end of the cold war, many governments, including Australia, have taken up the argument, traditionally associated with peace movements during the cold war, that states should now aspire to a nuclear weapons-free world, and aim to have zero nuclear warheads for all nuclear weapons states.

Nuclear issues are not just matters for the USA and Russia. The three other 'official', second-tier, nuclear states — Britain, France, and China — have not entered any negotiations for nuclear arms reductions since the end of the cold war, arguing in effect that the USA and Russia still have many more warheads than they do. Horizontal proliferation has also extended beyond these three to the 'unofficial' nuclear weapons states: Israel, India, and Pakistan.

Beyond these three, there is a group of states (such as North Korea, until recently) that are identified as being on the threshold of military nuclear acquisition, having both the technical capabilities and the possible motivation to develop nuclear weapons (the 'wannabes'). Offsetting this, some candidates have stepped back from the threshold, renouncing their intention to develop nuclear weapons (the 'coodabeens'); these include South Africa, Argentina, Brazil, and the three former Soviet Republics, apart from Russia, who inherited nuclear weapons from the USSR. The 'step back' is expressed by joining the Nuclear Non Proliferation Treaty (NPT) as a non-nuclear weapons state.

The NPT is an arms-control treaty initiated in the late 1960s by the cold war protagonists, the USA and the USSR, as an expression of their shared interest in containing the spread of nuclear weapons to other states. In return for others signing the treaty and renouncing the nuclear weapons option, the nuclear-armed signatories agreed to 'pursue negotiations in good faith on effective measures relating to cessation of the nuclear arms race at an early date and to nuclear disarmament', and to share the benefits of peaceful uses of nuclear energy with others. In 1995, a UN conference on the NPT agreed that the treaty should continue indefinitely. India, Pakistan, and Israel, however, have not signed the NPT.

The acquisition of nuclear weapons materials by non-state actors for possible 'terrorist' use has taken on a new post-cold war slant with a series of astonishing newspaper stories about the smuggling of plutonium and other nuclear-related materials out of the former USSR (*Washington Post*, 28 August 1994). 'Terror' may be a suitable description of the military policy of any state with nuclear weapons, but with a non-state actor, there is great difficulty in establishing a 'balance of terror' or 'deterrents', as such an actor has no territory to threaten in retaliation.

An added dimension of nuclear proliferation is the development of missile-delivery systems that operate over longer and longer distances, as states pursue the path taken by the USA and the USSR towards the intercontinental ballistic missile.

The NATO Secretary, General Willie Claes, summarised non-proliferation problems as he saw them in 1995 in the following way:

> One does not need to join the 'threat of the month club' to state that proliferation is bound to assume an ever greater significance in the coming years, for a number of technical and political reasons. Technically, the

ongoing debate about 'dual use' items indicate that the lines between civilian and military applications of a given technology are becoming increasingly blurred. Politically, the behaviour of Iran and North Korea has emerged as a major challenge to the non proliferation regime, as it stands for extension this Spring. Finally, the break up of the Soviet Union has added two entirely new — and totally unforeseen — dimensions to the proliferation problem: 'loose nukes' without political control or sufficient safeguards, and the 'brain drain' of nuclear expertise to other states.

(*Nuclear Proliferation News* 1995, issue 18)

Biological weapons, which also have the potential for enormous destruction, are prohibited by a UN convention that came into effect in the early 1970s. This treaty, unlike the more recently concluded Chemical Weapons Convention (CWC), has weak verification procedures, which are receiving new attention in the UN. Chemical weapons have been prohibited under a UN convention concluded in the early 1990s that came into effect in 1996.

The proliferation of 'conventional' weapons is also a major, but almost entirely unregulated, factor in global insecurity: high-yield conventional explosives and weapon systems may be as lethal as low-yield nuclear explosives. Furthermore, the wide distribution of certain 'low tech' weapons, such as land mines, can also have devastating consequences. The partially regulated 'legal' arms trade sits alongside an 'illegal' trade, making links between buyers with cash and sellers with weapons industries to support. The post-cold war world has been a 'buyers market', as past cold war players sell off some of their military hardware around the world. In the USA, some weapons lobbyists argue the need for more advanced weapons systems to be developed in the USA because they have sold such advanced weapons systems to other states in the system — arms racing with one's customers.

CONFLICT AND UN INTERVENTION

A second group of global insecurity issues arises from the patterns of conflict developing in the post-cold war world and the role of the UN in meeting, or failing to meet, the basic objective of removing the 'scourge of war' and maintaining 'international peace and security'. It has become commonplace to point out the rapid collapse of the collective security idea in the UN within a few years of its founding, in the face of the cold war and the increasing use of the veto in the Security Council. The UN developed a limited range of peacekeeping activities, for use in particular conflicts in which the parties had agreed to their role and the superpowers found the UN involvement useful. These peacekeeping interventions were usually referred to by their acronyms. For example, UNMOGIP (the United Nations Military Observer Group in India and Pakistan) was established in 1948 to assist the Indian and Pakistani cease-fire over Jammu and Kashmir; UNEF (the United Nations Emergency Force) played a similar role in the conflict between Egypt and British and French forces in 1956. In the 1960s, UN peacekeeping operations were dispatched to the Congo, West Irian, and Cyprus. In the 1970s, further peacekeeping forces were sent to the Middle East after the 1973 Arab–Israeli war.

As the cold war came to an end, the role of the UN expanded dramatically. Both the USA and the USSR/Russia sought to engage UN processes in resolving past cold war-fuelled conflicts in areas such as Namibia and Cambodia, and even in states once in their direct spheres of superpower interest, such as Nicaragua and Afghanistan. The removal of superpower fuel from the fires of many regional conflicts, and the re-engagement of the UN, gave some credence to the optimistic view of the management of post-cold war conflict patterns outlined in Chapter 1.

Iraq's invasion of Kuwait in 1990 represented an example of the kind of traditional cross-border aggression that the Charter of the United Nations was designed to deal with, and the massive response organised by the USA through the UN could be seen, at last, as the kind of collective security action envisaged by the UN's founders. Critics of the United States approach argued, however, that economic sanctions had not been given sufficient time to work. But whatever the arguments for and against the military response, the many exceptional dimensions of the Gulf conflict made it a precedent unlikely to be quickly emulated.

Cross-border aggression was the exception, not the rule, in the post-cold war world. Most wars were intrastate, and many of these were fuelled by ethno-nationalist claims, with one group often seeking greater autonomy, or even secession, from states in which they form a disaffected minority. These violent conflicts were no longer exacerbated by superpower rivalry, but nor were they controlled, as some once were, by superpower requirements of order within a sphere of influence. This is the 'pessimistic' scenario referred to in Chapter 1. The UN Charter gives considerable attention to the idea of non-interference in the domestic affairs of its member states, and many of the UN's member states have felt reluctant about endorsing UN interventions, as they may created precedents for interventions within their 'sovereign' jurisdiction.

But intrastate conflicts can rapidly develop profound interstate dimensions — through the creation of refugees seeking to cross into neighbouring states or leave by sea, through the temptation of external powers to back particular factions, and through the generation of serious humanitarian crises, which arouse the consciences of the citizens of other states.

In the first half of 1990s the UN was involved in some very difficult intrastate conflicts, with a mixed record of success. The successes in Namibia and Cambodia in the early 1990s were overshadowed by what are widely described as failures in Somalia and in the former Yugoslavia. In Somalia, the UN peacekeeping force entered in a blaze of media coverage, with the intention of alleviating the mass starvation caused, principally, by competing political factions in the country. Although the UN intervention was able to achieve some positive results in terms of feeding the population, it was unable to resolve the problems with the political structure, or gain the approval of the Somali political system for activities that went beyond peacekeeping and that involved more traditional use of military force ('peace enforcement'). Television footage of the body of a United States peacekeeper being dragged through the streets of the capital of Somalia, Mogadishu, spurred a reaction against UN interventions within the USA.

A similar dilemma for the UN was presented by the violent conflict in the former Yugoslavia. In Bosnia the UN role was initially limited to that of protecting safe havens against Bosnian Serb policies of 'ethnic cleansing'. But the UN role could appear to be a policy that in effect facilitated the very process it was attempting to prevent, in that it created ethnically based 'havens' that it was unable to secure from attack. In 1995–96, a United States-led NATO force intervened in Bosnia, effectively bypassing the UN, but securing its notional support.

Optimism about UN peacekeeping was brought sharply to an end by the Somali and Yugoslav crises, and following the uncertainties and hesitancy of the USA, the UN itself became uncertain and hesitant. In 1994–95 the UN failed to respond to the escalating ethnic massacres in Rwanda, allowing the slaughter of hundreds of thousands of Tutsis by Hutu militia. When the Hutu-led government was overthrown and the Hutu fled across the border, the UN further failed to prevent militia reorganising within the camps in preparation for the next attempt at genocide. The former president of the aid group Doctors Without Borders (Médecins Sans Frontiers) argued that in Rwanda what was needed was not 'peacekeeping', understood to be an activity that has the agreement of the conflicting parties, but 'peace enforcement', designed to prevent aggression of one party against another (Destexhe 1995). The failure of the UN to respond in Rwanda can be linked directly to the 'Somalia syndrome' in the USA, reinforced by a shift to the Right by the United States electorate — a shift that included a degree of 'anti-UN' sentiment on the part of key Republican leaders.

In 1995 the UN celebrated its fiftieth anniversary, and the future role of the UN in intervening to bring about global security had become contested and uncertain.

AUSTRALIA AND GLOBAL MILITARY SECURITY

The proliferation of deadly weapons and the development of deadly conflicts are clearly matters of fundamental importance to Australia's own long-term military security and to Australian citizens' sense of security.

Although these issues relate to long-term *military* security, they cannot be addressed by an Australian military defence policy designed to deter or repel an attack on Australia. Australia's military might pales against the collective military potential in the international system — as does the power of nearly all states. The point is not that these potentials are about to be collectively exercised against Australia, or that Australia is indefensible — far from it. The point is that the business of curbing the lethality of weapons systems and containing the conflicts that spur their development is essentially, for Australia, a diplomatic task, not a military task (Smith, G. 1992a). It is a diplomatic task that has generated roles for the Australian military, such as the verification of arms agreements and participation in peacekeeping. These roles have grown to the point that they are now explicitly discussed in defence white papers.

Neither smallness nor weakness has inhibited Australian governments in pursuing active diplomatic strategies aimed at improving global military security. Australian governments have acted as though they can 'make a difference'. But they have taken two largely divergent paths, based on two competing models — the realist and rationalist — through great power alliance and middle-power multilateralism respectively.

The realist approach to global security has sought to contribute to security by backing a great power ally. Australia's decision to join Britain in the two world wars, and then the USA in the cold war and its military conflicts in Korea and Vietnam, were in effect decisions about a model of global security as well as about Australia's place within it. The realist approach has focused more on the *conflict*, as a power struggle, than on *armaments*, which have been seen as a secondary dimension. According to the realist view, conflict has driven armaments acquisition rather than vice versa. Australian governments collaborated with equanimity in the development of British nuclear weapons in the 1950s, and accepted almost all nuclear weapon developments on the Western side of the cold war as necessary tools in the containment of communism. The termination of the cold war conflict that accompanied the collapse of the communist bloc could be regarded as a vindication of the realist approach, were it not for the massive stockpile of armaments inherited by the post-cold war world.

The rationalist agenda on global security has sought to build order-generating institutions and 'regimes' (such as arms-control regimes) through multilateral cooperation. Australia's role in the creation of the UN in the 1940s, in supporting arms-control development in the early 1970s (including taking French nuclear testing to the World Court), and its arms-control and UN-reform initiatives in the 1990s represent an alternative to global security through great power alliance. The rationalist approach often focuses on armaments, seeing weapons of mass destruction as a primary, and not a secondary, problem. Conflicts are generally seen as preventable, often as soluble via nonviolent means, and often as having been overtaken by the means (weapons) available to pursue them.

In the period of intense United States–Soviet competition in the last phase of the cold war in the 1980s, multilateral concepts struggled against alliance commitments, particularly in the approach of the Labor governments after 1983. Most of the time, multilateralism was not pursued at the expense or risk of the United States alliance. With the end of the cold war, the imperative of alliance politics seemed to ease, and rationalist agendas were given a significant impetus by the ALP's Foreign Minister, Senator Gareth Evans (see Evans & Grant

1995, ch. 6, for a description of Australia's diplomatic efforts in the pursuit of global security).

Weapons of mass destruction

Alliance politics led Australia to adopt a passive position on nuclear weapons in the 1950s and 60s, with Australia allowing territory to be used for British nuclear weapons tests. Australia relinquished its own right to acquire nuclear weapons by signing the NPT, as requested by the USA. In the late 1960s an Australian 'bomb lobby' of Liberal coalition backbenchers campaigned unsuccessfully for a government policy to develop an Australian nuclear weapon.

While renouncing its own nuclear weapons option, Australia participated in the transformation of ANZUS into a nuclear alliance with the establishment of United States nuclear weapons-related installations in the 1960s and 1970s. In the 'bases debate' of the 1980s (discussed in Chapter 3), Australia's Labor governments sought to defend the installations on global security grounds: as contributing to stability and reducing the threat of war. It was on these grounds (that is, instability and increasing the risk of war) that critics had called for the closure of bases or for Australian control. In contrast to New Zealand, Australian governments were not prepared to take domestic anti-nuclear measures that might jeopardise the alliance with the USA. Instead the ALP focused on a range of multilateral initiatives that were not seen to be in conflict with the United States alliance: the South Pacific Nuclear Free Zone Treaty, and in the UN, more vigorous participation in discussions on a nuclear test ban treaty, the extension of the NPT, and a very active role in the development of the CWC.

With the end of the cold war, several of these initiatives have emerged as major items on the international agenda. Australia made a large rhetorical shift and declared itself to be in support of a concept of a nuclear weapons-free world. Evans and Grant have stated that:

> As the number of nuclear warheads diminishes, international attention will turn increasingly to the ultimate target. The achievement of a nuclear weapon free world will require not only a political will, not now evident on the part of the nuclear weapon states, but also a major international effort to ensure that a total ban on nuclear weapons can be adequately verified and enforced.

> (Evans & Grant 1995, p. 87)

A key question is how actively and effectively Australia is pursuing its declared arms control and disarmament agenda. How is it managing its relationship with the USA in this process? After all, the USA still remains the most heavily armed power on the planet.

Comprehensive Test Ban Treaty

In 1963 the USA and the USSR signed the Partial Test Ban Treaty, partly as a result of increasing protest about the radioactive consequences of atmospheric nuclear test explosions. They subsequently conducted their test explosions underground. France continued to conduct atmospheric tests at Mururoa in French Polynesia in the Pacific Ocean, and in 1972 the Whitlam Labor government initiated proceedings in the World Court to ban French tests. The publicity surrounding this move persuaded France to shift to underground testing as well. A consistent theme of Australian arms-control diplomacy since 1972 has been to argue for the conclusion of a Comprehensive Test Ban Treaty (CTBT), which would ban all nuclear weapons test explosions whether atmospheric or underground. Such a treaty, it was argued, would put a cap on vertical proliferation, as it would restrain the modernisation of nuclear weapons. It would also inhibit horizontal proliferation, as would-be nuclear weapon states would not be able to test or perfect explosive devices. During the cold war the nuclear-armed powers, who were also the five permanent members of the UN Security Council, stalled the development of a CTBT.

With the end of the cold war, Australia has played an active role in the world's principal disarmament forum, the Conference on Disarmament (CD) in the UN, by proposing to move quickly to the conclusion of a CTBT. This has involved arguing and lobbying, against certain United States positions, on the mechanics and contents of such a treaty.

Nuclear Non Proliferation Treaty: indefinite extension

In May 1995 the members of the UN made a key decision to extend the operation of the NPT indefinitely. This indefinite extension was welcomed by Australia as offering 'by far the best encouragement for the nuclear weapon states to continue the historic process of nuclear arms reductions which had finally begun' (DFAT 1995). The five official nuclear-armed states supported indefinite extension of the treaty. Their critics argued that they did so because the treaty has lower expectations of nuclear weapon states than of non-nuclear weapon states. Indefinite extension requires that non-nuclear weapon states maintain that status, while not obliging the nuclear weapon states to adhere to a particular timetable for nuclear disarmament. Some non-nuclear weapon states had strong reservations about indefinite extension, fearing that it would remove pressure from the nuclear-armed powers to fulfil their part of the bargain under the NPT. Many non-government organisations were opposed to the idea of indefinite extensions for these reasons, and argued for a more limited extension and subsequent review in order to keep pressure on the nuclear-armed powers. In the Australian context these views were put strongly by, for example, Green Senator Dee Margetts (*Pacific Research*, vol. 8, no. 2, 1995).

Those with a cynical or sceptical view of the motives of the nuclear-armed powers found their views reinforced immediately after the permanent extension of the treaty. Within two days of the conclusion of the UN session, China conducted a nuclear test explosion, and within a matter of weeks the new French President, Jacques Chirac, announced that France also would be conducting a series of nuclear test explosions.

In the UN debate on this issue, Australia spoke out strongly for indefinite extension and so effectively became a backer of the United States position. To what extent was Australia's position influenced by its continuing alliance connections? Australia's independent position on nuclear issues was put to the test by the question of the legality of nuclear weapons. The International Court of Justice, or World Court, accepted a brief in 1995 to give an advisory ruling on the legality of nuclear weapons and invited submissions on this question. If the World Court were to declare nuclear weapons to be illegal, or if there were serious doubts about their legality because of their massive and indiscriminate killing capability, it would have ramifications for the global debate on nuclear weapons and may increase pressure on the nuclear-armed powers to take real steps towards nuclear disarmament. As a major nuclear weapons power, the USA supported their 'legality' and argued against a World Court ruling to the contrary. Initially Australia gave every indication of supporting the United States position in the World Court. But in a last minute change of position, Australia argued that, if the court must decide one way or the other, Australia endorsed the idea of 'illegality', although it would prefer that the Court make no judgement (DFAT 1996).

AUSTRALIA AND THE UN: PEACEKEEPER AND INSTITUTIONAL REFORMER

Catching the wave of optimism in the immediate aftermath of the cold war, Australia played a major role in the decision to establish a UN peacekeeping force in Cambodia to assist Cambodia in managing the transition to a democratic constitution following the end of the Vietnamese occupation (an occupation that had been justified by Vietnam on the grounds that it had evicted the murderous Khmer Rouge regime). Australia played a leading role in the UN Transitional Authority in Cambodia (UNTAC) by providing a significant number of personnel, particularly in the signals and communication area, and the military commander of UNTAC. Australia's role in UNTAC was its major military 'action' since the Vietnam War. The official perception of the mission was that it was a flawed success, but a success nevertheless (Smith 1994). By contrast, Australia's commitment to the Gulf War in 1990 appeared token, not just in terms of equipment or troops supplied, but also, of course, in terms of Australia's role in the overall scheme of the conflict.

The other major contribution that Australia made to the United Nations was that of Senator Evans, who, in the spirit of Dr Herbert Vere Evatt in the 1940s, produced substantial proposals for reforming the UN. These proposals were published in 1993 in a book titled *Cooperating For Peace*. In this work, Evans makes conceptual distinctions between UN roles in 'building peace', in 'maintaining peace' through particularly preventative diplomacy, and in 'restoring peace'. 'Restoring peace' is achieved through *peacemaking* (diplomatic methods used after a conflict has become an armed conflict), *peacekeeping* (the role of the UN developed in the 1960s and 1970s), and *sanctions and peace enforcement*, such as was used in the Gulf conflict. In recommending ways in which the UN could be more effective, Evans's emphasis is on the prospects of intervention at a very early stage in conflicts and on conflict prevention. Evans also had numerous suggestions about other aspects of UN reform, including the business of funding the UN, running an effective secretariat within the organisation, and promoting the idea of a UN standing military force.

Reviewers praised Evans's work as a useful and thoughtful way of thinking about the UN's role in peace and security. The Senator has encouraged speculation that he, like Evatt before him, is looking for a key position in the UN, possibly that of the Secretary-General (Evatt became President of the General Assembly in 1948). One review described his book as the longest job application ever written (Meaney 1991).

BACK HOME: DOMESTIC POLITICS AND GLOBAL SECURITY

Australia's global security diplomacy — in arms-control and disarmament negotiations in the UN and in seeking to reform the UN itself — is a game requiring advanced skill, talent and intellect. It appears to be an 'elite' activity requiring complex analysis and negotiation skills, and familiarity with the diplomatic culture of the UN. The mass peace movement that sprung to life in Europe and Australia in relation to nuclear disarmament concerns in the 1980s no longer exists in the 1990s. Until 1995 there seemed to be few connections made between Australia's global security diplomacy and its impact on domestic constituencies, apart from a few general assumptions about the views of the electorate, such as:

- the popular wish to rid the world of weapons of mass destruction
- the general desire for groups to live in peace with each other, especially as many Australians (or their parents or grandparents) have left regions that were marred by armed conflict
- the intuitive idea that a safer world means a safer Australia.

In 1995–96, however, domestic politics abruptly intervened in Australia's global security diplomacy.

From the CTBT to French tests

With President Chirac's announcement in 1995 that France would resume nuclear test explosions in the Pacific, there was a rapid upsurge of public resentment, which traversed both party lines and national boundaries — Australia's protests being echoed in New Zealand, the island Pacific states, many states of Europe, and even to some extent in Paris itself. In Australia the media reaction was spontaneous and widespread, with radio talk-back hosts from Left to Right, from commercial networks to the ABC, filling the air waves with anti-French-testing sentiment. Despite the absence of a large organised peace movement, there was a massive public reaction to the French announcement. (Some of the reaction expressed anti-French racism that was unable to distinguish between the nuclear policies of the French government and the aspirations of the French people.) Nevertheless, the extent of the reaction suggested far-reaching public expectations that, in the post-cold war world, nuclear-armed powers would engage in new forms of international behaviour. The Australian Foreign Minister appeared to be caught off guard. Senator Evans's initial response, made in Japan, was couched in the language of state-to-state multilateral negotiations: what the French were doing was regrettable, but we should take note that there was only a limited testing program proposed, at the conclusion of which we hoped the French would join the rest of the world and sign a CTBT; we could look forward to these being the last French nuclear tests. The domestic response in Australia was one of more straightforward anger and, in effect, an enthusiasm for a kind of people-to-people transnational diplomacy to operate against French policy. Senator Evans was widely seen as being too timid, as giving the political initiative on the issue to the Liberal Opposition, and even as having damaged his political career.

From reforming the UN to East Timorese self-determination

In July 1995 a conference to celebrate the fiftieth anniversary of the UN was held at La Trobe University in Melbourne over five days. The conference was opened by Senator Evans — a tribute to his intellectual and political commitment to UN reform. The Labor Foreign Minister's speech emphasised the themes of his book, focusing on preventative diplomacy and on peace-building, particularly in light of those intractable intrastate ethno-nationalist conflicts that beset many parts of the world. But with one of these ethno-nationalist conflicts close to Australia, that between the East Timorese and the Indonesian government, the failure of preventive diplomacy was not only manifest, but had even transformed itself into a crisis in Australian–Indonesian relations. At the time of the conference, the Foreign Minister was facing vocal public protest in response to an Indonesian proposal, apparently accepted by Australia, to send General Mantiri to be Indonesia's next Ambassador to Australia. General Mantiri had made comments endorsing Indonesian behaviour in the massacre of East Timorese at Dili in 1991.

Inside the conference, 500 delegates from Australia and around the world listened to the Foreign Minister discuss the principles of preventive diplomacy. Outside the conference walls, there were 200 protesters, sixty police, and enough noise to dominate conference proceedings. The message from outside the conference hall was that Australia was not taking a sufficiently principled approach on the issue of East Timor. Inside the conference venue the same question was being asked in a somewhat more elaborate way. What was Australia doing to utilise the UN, and to seek to apply the principles of preventive diplomacy and conflict resolution in East Timor? *Age* journalist Michelle Grattan reported that, as a result of the experience of the protest at the conference, Senator Evans set out to stop the appointment of the Indonesian ambassador, at some cost to Australia's diplomatic standing with Indonesia (1995).

Domestic pressures and global security diplomacy

Domestic pressures do not necessarily conflict with global security diplomacy. Indeed, domestic pressures may strengthen global security diplomacy by giving the state a push from below. One risk of multilateral diplomatic activity is that diplomatic culture can be seductive and it is easy to develop a sense of complacency about incremental progress, which may not be shared by domestic constituencies, who notice instead the gap between the size of the problems and that of the international community's response.

These intrusions of domestic politics into global security issues seemed to have had some effect on Australian policy-making. The public reaction in Australia to French nuclear tests may have served to toughen Australia's position in negotiating forums on disarmament. For example, it appears to have contributed to a stronger anti-nuclear weapons position being presented to the World Court on the question of the 'legality' of nuclear weapons than the expected Australian justification of the nuclear weapons of our United States ally. On the question of ethno-nationalist conflict, domestic pressure may have encouraged greater attention to a specific conflict than is afforded by generalised schemes for dealing with a whole class of conflicts, encouraging a more vigorous approach to the realisation of East Timorese self-determination, with less concern for the non-negotiable claims of Indonesian sovereignty. These are conjectures, but the events of the mid-1990s may foreshadow a much greater interaction between domestic politics and global security policies throughout the rest of the decade.

CONCLUSION

Global security concerns, by their very magnitude and intractability, can be placed by Australian governments in the 'too hard' basket, and can be accepted as part of the landscape of international politics that Australia is unable to change, even if under a degree of internal pressure to do so. Great powers, including allies, do not necessarily welcome the intrusion of smaller powers into issues affecting their approach to armaments and conflict.

In (rationalist) theory, middle powers, by forming coalitions, can play a key role in addressing issues that immobilise the larger powers. The extent to which Australian governments will put intellectual energy, political skill, and financial resources into meeting the challenges of global military insecurity depends on the extent to which they are able to continue to conceive a role for Australia that goes beyond realism or fatalism: the role of an effective, order-creating middle power, working with other 'like-minded' powers on armaments and conflict issues.

7   Regional Military Security

INTRODUCTION

'Regional military security' refers to a state of affairs in which not only is Australia at peace with its neighbours, but military threats from neighbours are also very improbable and high level threats almost inconceivable. Australian governments see maximising regional security as a major objective, and this objective is often expressed as a desire to foster a benign or favourable strategic environment. Military security is not an absolute, but rather a matter of degree, and realists are not the only ones to acknowledge a fundamental element of insecurity in an anarchic international system, no matter how favourable a state's geopolitical location. The problem of *regional* military insecurity, like that of *global* military insecurity, arises from the operation of the international system as a threat system. But regional security is concerned with those threat potentials closer to Australia.

Proximity has two implications. First, the capacity to project military power is fundamentally affected by distance: a military force attacking over a longer distance faces far greater logistical and resupply problems than one attacking over a shorter distance and is potentially more vulnerable to interdiction. A direct 'conventional' threat from a state in the region would be less difficult and expensive to mount than a threat from a state further away. This is not to say that neighbours must fight, that they are likely to, or that they ever will fight, but is simply an acknowledgment that it is generally easier for a military attack to operate over shorter distances.

The second implication of proximity is the converse: a conventional military *defence* is also much easier to mount over shorter distances from Australia than longer ones. Planning for military defence may be part of the response to regional military insecurity, and part of the foundation of regional military security. (A defence against a global threat such as a nuclear-armed ICBM is almost inconceivable for Australia.) An integrated Australian regional security approach will need to address the relationship between foreign and defence policy. A successful foreign policy is one that creates a regional environment in which military defence is not used. On the other hand, defence forces can be seen as insurance against an unsuccessful regional foreign policy.

Over the last century there have been considerable shifts in official and popular perceptions of regional insecurity, and changing threat perceptions have been discussed in Chapter 2, on the making of foreign policy, and chapter 3, on the ANZUS alliance.

Paul Dibb, as adviser to the Minister for Defence, wrote in 1986 that 'Australia's neighbours possess only limited capabilities to project military power against it ... Australia faces no identifiable direct military threat and there is every prospect that our favourable security circumstances will continue' (Dibb 1986, p. 1). In 1995, as Professor Dibb, he put a more pessimistic slant on his observations:

> Economic growth will increase the military power of the major local players [in Asia] and political change may make their policies less predictable. Economic interaction can be a stabilising force, but it can also intensify frictions over policy and security issues. In view of these uncertainties, it is possible that Asia's security could deteriorate, perhaps quite seriously, in the future.
>
> (Dibb 1995, p. 16)

As with all reflections on international politics, we can ask how much this assessment reflects the realities of the world 'out there' and how much it is a product of the mind set (or vested interests) of the writer. If, for the moment, we leave aside the question of how threatening or non-threatening the regional state system is, we can first clarify a few points of definition, without making any judgement on how secure or insecure Australia may be.

Regional military *insecurity* can refer to a spectrum of hypothetical situations, the severity of which depends on the probability and magnitude of military threat. In terms of probability, the state of the regional environment may range from one in which threats are highly probable (or, at the extreme, where armed conflict exists) through to situations in which they are less probable, to circumstances in which threats may still be possible but very improbable. In order of declining magnitude, military assault can range from invasion, to large-scale attack, to small-scale attack, to a 'lodgement' or hostile action by a single member of another state's armed forces.

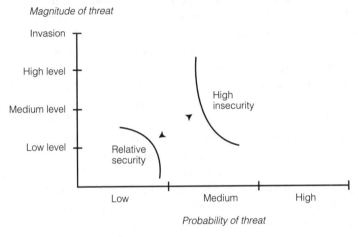

Figure 7.1 The relationship between security and insecurity

Figure 7.1 shows the probability and magnitude of threats on two axes. The area in the top right corner of the graph represents a situation of extreme insecurity, characterised by high-probability, high-level threats; the area in the bottom left corner represents relative regional security, characterised by low-probability, low-level threats.

A regional security policy is one aimed at minimising the probability and magnitude of military threats, with the basic objective of providing as much freedom from threat as will allow the Australian community to set its own goals and priorities through democratic processes without military pressure, or a fear of military pressure, from without.

EARLY APPROACHES TO REGIONAL MILITARY SECURITY

Australia's approach to regional security has moved through several historical phases since federation, parallelling the broader ideas of security and region that have prevailed at different times.

Britain and Empire

The period from federation in 1901 to the Second World War saw the continuation of the colonial pattern of dependence on Britain as

Australia's security guarantor. Britain, a global power, had major naval bases in its colonies in Singapore in South-East Asia, Hong Kong in North-East Asia, as well as control of all of what are now India, Pakistan, Bangladesh, and Burma. This dependence fundamentally affected Australia's outlook on the Pacific and Asian regions.

Australia's approach to the Pacific islands — that of strategic denial, or preventing potentially hostile powers from securing possible bases — reflected the concerns it had expressed in the nineteenth century. Australia urged Britain to pre-empt other powers by asserting sovereignty over island groups and, increasingly, sought control itself. Britain passed the administration of Papua to Australia in 1906. With the outbreak of the First World War, Australia took control of the German territories of New Guinea and Nauru, and retained control after the war with the endorsement of the League of Nations. Australia's approach to the Pacific islands in this period was likened, by Australian prime ministers, to the United States Monroe Doctrine, which sought to deny Central and South America to any potentially hostile power (Fry 1991, ch. 1).

As for what came to be called the 'Near North', Australia's showed considerable indifference towards Asia (as discussed in Chapter 4). Anxiety about Japan's aggression in China in the 1930s was alleviated by the presence of British forces in Singapore. The extent to which Australia's security was oriented towards 'Empire' was evident at the outbreak of the Second World War in Europe in 1939, with the dispatch of Australian troops to Europe and the Middle East, following the pattern of the First World War. Britain would look after the region, and Australia would help Britain look after the world.

Australia's military forces in peacetime were developed along the dual lines of a British Empire-oriented career-officer class, along with a reliance on volunteer and conscripted militias to train in various aspects of home defence. These armed forces were not intended to have any particular impact on regional security.

Alliance to 1972

Japan's entry into the Second World War in December 1941 remains the most serious regional security crisis in Australia's history. Australia was threatened with invasion by a regional power, and the threat seemed imminent. Air attacks were carried out on northern towns, and Australians were killed on their own territory. Australia's immediate response was to seek to divert some of its military resources away from Britain's strategic control to the more immediate task of defending Australia, leading to Prime Minister John Curtin's celebrated clash with Winston Churchill when he insisted that troops return directly to Australia.

One reading of the lessons of the Second World War for Australia's regional security policy is that Australia had been so reliant on a great power that it had neglected to take responsibility for its own military defence against possible regional threats. A more regionally attuned, self-reliant foreign and defence policy is the solution suggested by this analysis.

But an alternative set of lessons was learnt. Although Britain had failed to protect Australia, the world's greatest power had saved the day. If dependence on Britain could be transferred to the USA, Australia would remain protected by a great power from regional threats. And so began the pursuit of the United States alliance.

The cold war ideological classification of states in the region as either communist adversaries, anti-communist allies, or those who were non-aligned and manipulated has been discussed in Chapter 4. This classification was part of the Liberal government outlook from 1949 to 1972. The Australian variant of the United States and British worldview held to the pervasive image of an overwhelming Chinese communist threat. When Robert Menzies announced the dispatch of troops to the Vietnam War in 1965, he announced in Parliament that the basic cause of the conflict was the downward 'thrust by Communist China between the Indian and Pacific Oceans' (House of Representatives, *Debates*, 29 April 1965, p. 1061). A massive Chinese communist threat against an indefensible Australia justified the United States alliance and participation in war with the USA.

Fear of communism or fear of 'Asia'?

To what extent was Australia's cold war ideological perspective on the region a projection of more deep-seated fears of 'Asia', irrespective of the communist or anti-communist character of Asian societies? Menzies expressed public regret that European colonial powers might depart from the region, and painted a picture of regional instability and threat to follow the loss of European control. European control was seen as less threatening than independent Asian states, whatever their political complexion. The independence of Indonesia, he thought, would unravel the colonial system in Asia, leading eventually to Australia's disintegration (Lowe 1994). To what extent were Australians still operating with nineteenth-century cultural images of the 'Yellow Peril' rather than the cold war political image of communist China as the 'Red Peril'?

There was also a curious resonance between China's so-called 'downward thrust' and Japan's actual military strategy in the Second World War, as though the past behaviour of Japan had been transposed onto China. In the 1950s academics debated about whether 'Asia' was a 'threat to Australia', as though Asia was a state, or an entity that could somehow act collectively against Australia (Broinowski 1992).

Derivative regional involvements

Australia's involvements in wars and treaties in the region in this period derived from alliance arrangements, and followed the lead of Britain and the USA. These involvements were subordinated to the overall purpose of alliance politics: to secure the support of the alliance partner against a future threat from the region.

Australia became involved in several treaty arrangements involving states in the region. The Philippines was a member of SEATO, and the non-communist states of Indo-China were covered by its protocols. Australia also became a party to the Five Power Defence Arrangements with Malaysia and Singapore, which grew out of the earlier ANZAM arrangements established to facilitate British withdrawal from its Malayan colonies. But these treaty arrangements were sponsored by great powers — the USA and Britain, and Australia's principal alliance ties remained with these powers. In the 1950s and 1960s Australia was not involved in any bilateral military security arrangements with any Asian state or in any multilateral arrangements that did not include Britain or the USA.

After the Second World War Australia made a fundamental change in its defence policy: it moved to the maintenance of a standing professional armed force, rather than a reliance on militias or part-time soldiers. But there was also a fundamental continuity: defence forces were developed essentially to be the adjunct of a great power, for use in 'forward defence' rather than for the defence of Australian territory. Any role in regional security was incidental to this alliance role. Australian troops would be engaged in the region if that was where Britain or the USA sought to be engaged. These great powers would in turn protect Australia from regional threats.

REGIONAL SECURITY: REASSESSMENTS

Four reassessments of Australian regional security have taken place since the heyday of alliance politics in the 1950s and 1960s. The first two have responded to the *rapprochement* between the USA and China in the early 1970s. The Whitlam Labor governments, and then the Fraser Liberal governments, sought to accommodate the massive shift in United States policy towards China, and China's implacable opposition to what it saw as Soviet expansionism. The second two reassessments have been largely in response to the United States–Soviet thaw of the mid-1980s, and the subsequent end of the cold war and collapse of the USSR. In the 1980s and 1990s, government ministers struggled to redefine Australia's regional security role in a region in which Soviet power had evaporated and United States power was contracting. Kim Beazley's 'new militarism' was supplanted by Foreign Minister Gareth Evans's commitment to 'security as part of the region'.

1 Whitlam 1972–75: diplomatic engagement

Prime Minister Gough Whitlam declared that Australia's foreign policy would take a new direction under his leadership, improving Australia's 'standing' in the region: 'Our thinking is towards a more independent Australian stance in international affairs which will be less militarily oriented and not open to suggestions of racism; an Australia which will enjoy a growing standing as a distinctive, tolerant, cooperative and well regarded nation not only in the Asian and Pacific region but in the world at large' (Whitlam 1973). An Australian policy of greater regional orientation was given considerable impetus by the United States–Chinese *rapprochement*, symbolised by President Nixon's visit to Beijing in 1972. The ALP felt a sense of vindication as its long-espoused regional policy of the recognition of communist China was now fully in line with United States policy rather than directly at odds with it.

Whitlam declared Australia would like a new regional organisation to be formed to discuss shared security problems, and he set off on a tour of the capitals of Asia in an attempt to persuade Japan, China, and India to consider the idea. The proposal was abortive, but prefigured the successful institution-building in the region that was to be a hallmark of the later Hawke and Keating governments.

Whitlam declared an end to the era of forward defence and, in addition to withdrawing the remaining troops from Vietnam (along with the withdrawing United States forces), foreshadowed the with-drawal of Australian troops from the Malaysian base at Butterworth, where Australia maintained a significant airforce presence. The Minister for Defence, Lance Barnard, said there was no foreseeable threat to Australia from the region in the next fifteen years and initi-ated the first concerted attempt to formulate an appropriate Australian defence policy for such a relatively secure regional environment.

2 Fraser 1975–83: anti-Soviet geopolitics

The United States *rapprochement* with China was initiated by President Nixon, principally because it was thought it might assist the extrication of the USA from the Vietnam War by leading to Chinese pressure being put on the North Vietnamese to settle for a partitioned country. For the Chinese, the *rapprochement* with the USA was seen principally as part of an anti-Soviet strategy, as China had become convinced, since its fallout with its former ally, that the USSR was now the major threat to both China and the world. Understandably, then, the United States–Soviet *détente* alarmed the Chinese.

From 1978 on, the United States–Soviet *détente* collapsed, and United States–Soviet relations deteriorated into the 'second cold war'. The United States–Chinese relationship then took on more directly geopolitical overtones as a de facto alliance against perceived Soviet expansionism. It was this kind of geopolitical worldview that Malcolm Fraser found attractive, and it allowed the Liberals, after more than twenty years of anti-Chinese-communism rhetoric, to reinvent China as a staunch ally against the greater evil of Soviet communism.

Fraser was thus able to continue the Labor policy of accord with China, and to go further to establish even warmer relations with China than those under Whitlam. This warmth reached the point at which the Australian Prime Minister made announcements congratu-lating the Chinese on the success of ballistic missile tests in the Pacific.

Whereas anti-communism had limited earlier Liberal govern-ments' positive regional relationships (such as those established by the Colombo Plan) to acceptable anti-communist states in the region, the new, more-focused anti-Sovietism allowed some wider connections within Asia. However, as these regional relationships were driven by a geopolitical strategy, they remained limited by the narrowness of this approach. This is most clearly illustrated by Australia's response to events in Cambodia. The extensive killings by the Khmer Rouge government in the period 1975–78 were stopped by a Vietnamese invasion. A Vietnamese-installed government was placed in Phnom Penh. But because Vietnam had formed an alliance with the USSR, the Fraser government felt obliged to support the coalition of resistance groups attempting to overthrow the new regime in Phnom Penh. Part of the coalition was the deposed Khmer Rouge group.

Fraser's diplomacy also developed a non-regional focus on African conflicts, where he was instrumental in the removal of the minority regime in Rhodesia and the creation of the state of Zimbabwe.

3 Beazley 1986–89: new militarism?

The Hawke Labor government was elected at the height of the 'second cold war', and as a consequence, global security issues rather than regional security concerns were at the forefront of foreign policy for several years.

As a new period of United States–Soviet *détente* developed, the Australian Minister for Defence, Kim Beazley, set out the government's new defence policy in the Defence White Paper of 1987. On one level the White Paper represented a fundamental shift from forward defence concepts towards the defence of Australia, setting out the kind of military threats Australia might have to face, how the Australian military forces would respond, and what kind of equipment, personnel, and organisation would be required for effective response. This could be seen as a shift to a 'normal' defence policy, away from one premised on a belief in the indefensibility of Australia. This development in defence policy was exceptional only in that it took so long to be articulated. It was the kind of policy that Australia may have developed after the Second World War if it had not sought protection though military alliances.

However, the content and presentation of the new defence policy raised key issues relating to Australia's regional relationships. In introducing the White Paper, Beazley declared:

> This program is the largest defence capital investment in Australia's peacetime history …
>
> The strategy on which self-reliance is based establishes an extensive zone of direct military interest …
>
> Australia's long-range strike capabilities are being developed to respond — quickly and lethally — to early warnings far from Australia's shores …
>
> Australia's surface and submarine naval forces are undergoing the most

dramatic expansion in peacetime history ...

Long-range ships, submarines and aircraft, and highly mobile ground forces, will enable us to play our proper role in the region, and, if necessary, beyond it.

(Department of Defence 1987)

Taking Australian military defence seriously for the first time also meant taking the regional implications of defence policy seriously, and it was alleged shortcomings in this area that were highlighted by critics who saw a 'new militarism' in Australia's regional security approach (Cheeseman & Kettle 1990; Sullivan 1995; Dalby 1996). Journalist David Jenkins has written of the domination of foreign policy by defence policy under the influence of 'Bomber Beazley' (Jenkins 1989a; 1989 b).

In September 1989 a group of sixty-eight prominent Australians, including Sir Mark Oliphant, Janine Haines, Bob Brown, and Peter Garrett, signed a Statement of Concern that began:

Australian military policy is entering a costly and dangerous new phase, in the almost total absence of public debate.

The Australian Government, supported by the Opposition parties, is strengthening the ability of our military forces to attack our South-east Asian neighbours, and has mobilised troops to intervene in island states of the South-west Pacific during recent upheavals. Australia's peace-time military presence in the region is being stepped up.

(Cheeseman & Kettle 1990, pp. 207–8)

This depiction and critique of Australia's military policies has been described as the new militarism thesis. The arguments outlining and criticising the 'new militarism' in Australia were developed in 1989–90 by the Secure Australia Project, and can be summarised in six points:

1   **Build up of long-range weapons systems**. The popular image that Australia has insignificant military power and was 'not able to protect Botany Bay ... on a hot Sunday afternoon' is a relic of more self-deprecating days. Australia has, and is fast developing, potent military capabilities: long-range hi-tech weapons systems with multi-billion dollar price tags — 75 state-of-the-art FA-18 fighters, an avionics upgrade of its existing 22 F-111 bombers, the construction of eight 'ANZAC' frigates, new destroyers, six to eight new submarines (said to be the 'most powerful conventional submarines in the world'), and more.

2 **Alarmism — spinning threats.** Superpower arms reductions have been largely ignored in government statements in favour of an image of dynamic regional powers who may threaten Australia in some undefined military way for some unknown reason some time in the future. 'Instability' in the South Pacific has been discussed as creating circumstances in which Australian military intervention may be justified.

3 **Regional 'role' — the new forward defence.** A new willingness to consider dispatching an Australian expeditionary force to regional trouble spots was evident in the mobilisation of Australian troops during the Fiji coups in 1987 and the Vanuatu riots in 1988. Routine deployments of Australian naval and air forces in the South-east Asian and South Pacific regions had increased. Australia's area of 'direct military interest', in which our military force will prevail, is set at an arbitrary 1000 miles — far from Australian territory and offshore economic zone.

4 **Regional Consequences.** A New Zealand Labour Party foreign affairs committee has described Australia as the most destabilising power in the South Pacific. Intense Australian pressure on the New Zealand government achieved a New Zealand decision to purchase 'ANZAC' frigates, a decision opposed in opinion polls by over 70 per cent of the New Zealand population. Bougainville secessionists accused the Australian government of providing light arms and helicopters used in the PNG army's unsuccessful bid for military victory in 1989. While the Indonesian government has declared itself relaxed about Australian military developments, a number of prominent Indonesians have expressed concern over the new Australian militarism (Eccleston 1989). Arms racing can still proceed in an atmosphere of polite diplomatics.

5 **Economic and Social Dimensions.** An indigenous arms industry is being fostered through directing a substantial part of the equipment budget to local procurement, funding the Defence Science and Technology Organisation to the tune of $200 million per year, and encouraging arms exports under relaxed guidelines …

6 **Security Consequences.** By putting so many eggs in the long-range high potency weapons basket, Australia lacked basic capabilities in coastal surveillance to counter the low level but potentially serious threats of quarantine violation, poaching in the off-shore economic zone, illegal immigration and drug smuggling. Furthermore by failing

to develop adequate capabilities for very low level military conflict (such as harassment of coastal shipping), the government may be forced into escalation in a crisis. Developing long-range offensive capabilities may, of course, encourage others to do likewise, so diminishing our security.

(Smith 1992b)

4   Evans 1989–95: security as part of region?

Gareth Evans was appointed foreign minister in 1988, and his predecessor, Bill Hayden, moved on to become Governor-General. Evans provided a significant intellectual counterweight to the 'new militarism' and steadily shifted Australia's regional security discourse into a more complex and multidimensional framework. Kim Beazley moved on from Defence to other ministerial positions, and his successor, Robert Ray, did not seek a high intellectual profile in defining Australia's role in the region.

Evans's first systematic attempt at theorising was his 1989 ministerial statement *Australia's Regional Security* (as quoted in Fry 1991), which sought to identify the wider range of factors that contributed to Australia's military security apart from military preparedness:

The protection of Australia's security ... has more than one dimension. In specifically military terms, it involves the capability to deter, and if necessary defeat, attacks against us; in broader external policy terms, it involves acting, through many different policy instruments, to maintain a positive security and strategic environment in our own region.

The instruments of policy required to protect our security interests go well beyond those administered by the Minister of Defence, or for that matter the Minister for Foreign Affairs and Trade ... they may be identified as follows:

1. Military Capability;
2. Politico-Military capability;
3. Diplomacy;
4. Economic Links;
5. Development Assistance;
6. 'Non-Military Threat Assistance; and
7. Exchanges of People and Ideas.

(Evans, quoted in Fry 1991, p. 184)

The statement distinguished between relations with South-East Asia, on the one hand, and the South Pacific, on the other. The approach to South-East Asia was 'comprehensive engagement', and the approach to the island Pacific was 'constructive commitment'.

According to those who had criticised the 'new militarism', Evans made too many concessions in his approach to the significance of military power in Australia's regional security policy, with his attraction to concepts such as 'politico-military capability' (Cheesman & Bruce 1996). But the statement represented a successful bid by the Foreign Minister to take control of the larger regional security agenda, and the rapid pace of events confirmed and extended its 'multidimensional' thrust.

The end of the cold war in Europe led to dramatic changes in South-East Asia as Vietnam, without a Soviet ally, agreed to withdraw from Cambodia and began to normalise its relations with ASEAN. On the other side, the Western powers were able to break off their association with the Khmer Rouge and convince China to do likewise. The spectre of Soviet penetration of the island Pacific vanished. The USA closed down its massive naval and air bases in the Philippines.

Evans moved steadily to develop regional security thinking in light of the dramatically changing external circumstances. He outlined a set of ideas to guide Australia and the region's approach to military security. These ideas included:

- 'common security', whereby states seek to establish win–win military arrangements, along with confidence-building measures that are integral to any national defence effort, rather than just being an afterthought. Not threatening others and creating a cycle of threat is as important a principle as protecting oneself.
- 'comprehensive security', whereby the non-military dimensions (for example, environmental and economic dimensions) are seen as significant in their own right and not just to be considered because of their connection to possible military threats.
- 'human security', in which the focus is on the security of the individual, not just that of the state.
- 'cooperative security', used as an overarching term to describe the combination of the above ideas.

This expanded vocabulary of regional security was combined with initiatives to establish institutional processes in the region that could give these ideas some expression (Ball & Kerr 1996).

The first concrete initiative was Australia's promotion of UN involvement in attempts to create a new political structure for Cambodia and to put an end to armed conflict. Australia subsequently made a major contribution to the UN Transitional Authority in Cambodia (UNTAC) by providing the commander for the operation and over 900 troops, who were mainly used to establish the communications system.

Another major multilateral initiative was to promote a new regional security dialogue. When Evans first proposed an institution in East Asia that would be similar to the Conference of Security and Cooperation in Europe, he was criticised for attempting to apply to Asia a formula that was specific to Europe. But the proposal subsequently received support from President Clinton, and the formation of the ASEAN Regional Forum in 1994 owed much to Australia's efforts.

Evans summed up the process of change that had taken place in Australia as a shift from seeing Australia's military security as being 'against the region' to seeking security 'as part of the region'. This required a complex multifaceted regional diplomacy in place of a dependency on a great power from outside the region.

FIVE REGIONS AND THEIR MILITARY SECURITY IMPLICATIONS

What is Australia's region? This apparently simple question that has no simple answer was discussed in Chapter 4, where it was observed that the notions of 'our region' or 'being part of a region' are politically constructed ideas, designed to foster a sense of solidarity with other states for economic, political, or cultural purposes.

When considering regional military security, the discussion is sometimes simplified along the lines of a twofold division between:

- the 'North', or 'Asia', where potential threats were identified as the European powers retreated, and
- the Pacific islands, through which it was believed threats might be exercised.

This is too great a simplification, and when considering the issue of regional military security, a more conventional geographical approach is possible. Such an approach examines all of Australia's 'regions' without attempting to define what is 'our' region. Starting to the south of the Australian continent and sweeping in an anticlockwise direction, five 'regions' can be identified, and each can be analysed in terms of Australia's military security (that is, from the perspective of potential military threats and of patterns of conflict that may generate such threats). The regions are:

1 Antarctica
2 The island Pacific, or Oceania
3 South-East Asia (the ASEAN states, including Indo-China — Vietnam, Cambodia, Laos)
4 North-East Asia (China, Taiwan, Japan, and the Koreas) and
5 The Indian Ocean, including the states of South Asia (India, Pakistan, and Bangladesh).

1 Antarctica

Antarctica, the vast ice-covered continent situated across the Southern Ocean from Australia, has never been seen as the source of any kind of military threat to Australia. A snapshot of human activity in Antarctica today would show a series of scientific enterprises run by about twenty states, with a summer population of about 4000, a winter population of about 1000, and a very low level of tourism. Any question of military potential is a speculation on future developments and might be dismissed as fanciful were it not for the fact that two of the classical ingredients for interstate conflict are present in Antarctica.

First there is the political map. Antarctica is territorially carved up like a cake into wedges that meet at the South Pole. The slices are claimed as sovereign territory by Australia, New Zealand, France, Norway, Britain, Argentina, and Chile. Some of the slices overlap and sovereignty is contested — Chile overlaps with Argentina, and both overlap with Britain. Other states contest the sovereignty of all the claimants, arguing that none of the claims should be recognised. Second, there is, for some, the alluring prospect of Antarctic mineral resources. The land under the ice may contain an abundance of minerals, and prospective ocean resources exist within what would elsewhere be 200 nautical mile Exclusive Economic Zones. The political map suggests the possible scenario of old-fashioned disputes over territory leading to a militarisation of Antarctica; the prospect of resource extraction raises the stakes in any such potential disputes.

Innovative diplomatic and international legal solutions to these two issues have been found, solutions that mean Antarctica will remain an improbable source of military threat in the future. The Antarctic Treaty, concluded in December 1959 by twelve states, including cold war antagonists the USA and the USSR, put the sovereignty claims 'on ice'. The treaty established Antarctica as a military-free zone. Article 1.1 of the treaty states:

> Antarctica shall be used for peaceful purposes only. There shall be prohibited, inter alia, any measure of a military nature, such as the establishment of military bases and fortifications, the carrying out of military manoeuvres, as well as the testing of any type of weapon.

Over forty states are now consultative parties to the treaty. The Protocol on Environmental Protection, adopted in 1991, effectively banned mining for fifty years because of the ecological fragility of Antarctica and made it difficult for mining proposals to be accepted after that date. Antarctica was designated a 'natural reserve dedicated to peace and science'. Australia, in conjunction with France, played a significant role in securing this 'world park', although not until the Hawke Labor government abandoned its support for a Minerals Convention, which would have allowed the possibility of regulated mining. Australia's role here exhibits an intriguing combination of reliance on non-government organisations (for the basic form of the protocol) and vigorous multilateral diplomacy.

Potentially undermining the Antarctic Treaty system is the fact that it has not been concluded directly under UN auspices, and some states have argued that all the arrangements should be redrawn under the direction of the UN. Malaysia, in arguing such a case, has emphasised the mineral resource potential of Antarctica and the injustice of denying resources to a resource-scarce world.

2 Oceania

Oceania consists of about twenty island states and territories, apart from Australia. Papua New Guinea and New Zealand, each with 3.5 million people, are followed by Fiji, with 700 000, and the Solomon Islands, with 250 000. The populations of New Caledonia, Vanuatu, Western Samoa, and French Polynesia are between 100 000 and 200 000. The rest are truly microstates, both in terms of population and land area.

The peoples of the islands are often loosely classified as either Melanesian (Papua New Guinea, the Solomon Islands, Vanuatu, and New Caledonia), Micronesian (the islands nearer the equator, running from Palau in the east to the Marshall Islands in the west), or Polynesian (the rest). But there have historically been many more language and cultural groups than this classification suggests. Furthermore, in Fiji, about half the population are descended from Indian labourers brought over by the British, and in New Caledonia, about 40 per cent of the population are French settlers or descendants of French settlers.

The South Pacific Forum is the major political organisation that defines Oceania, and the region is also loosely described as the 'South Pacific', or simply the island Pacific, to distinguish it from the Pacific Rim states. There are fifteen self-governing member states, several in free association with a larger power.

The most obvious strategic fact about Oceania is that the island states, individually or collectively, cannot pose a military threat to Australia. Today only Papua New Guinea, Fiji, and the Solomon islands have military forces, and these have no power-projection capability.

The strategic concern that has been intermittently expressed in Australia for over 100 years arises from this very weakness of the islands: a concern about possible threat not *from* the island Pacific, but *through* the island Pacific. The fear has been that a hostile external power may establish bases in the islands either to threaten Australia directly or to threaten its sea lines of communication.

Colonial governments of the 1850s worried about European powers such as France and, later, Russia and Germany; the Australian governments of the 1980s worried about Russia again (in the form of the USSR). The Australian response for over a century has been a strategy of *denial*: an attempt to prevent access or influence by the unwanted external power or powers of the day. Denial could be achieved by involving an ally, such as Britain, the USA, or New Zealand, or by unilateral Australian action. At the end of the Second World War, for instance, Australia supported the retention of United States control over the former Japanese-ruled islands in Micronesia, and the British also had a long history of involvement in Australia's policy of strategic denial.

In the 1960s and 1970s the global decolonisation movements extended to the Pacific. Britain, Australia, New Zealand, and eventually the USA allowed island states to have political independence or 'independence in free association'. (France, however, retained control of its territories.) Western Samoa became independent in 1962, Fiji in 1970, and Papua New Guinea in 1975. Independence created new problems for the strategy of denial: island governments were now potentially independent decision-makers who might wish to enter into relationships that Australia saw as having military-threat potential. It was a matter not only of pre-empting or forestalling external powers, but also of persuading a new set of actors of the validity of Australia's regional security perceptions.

From the late 1970s to the late 1980s, Australian governments under Fraser and Hawke declared their anxiety about the prospect of Soviet penetration of the island Pacific. In particular, concern was expressed that fishing agreements concluded between island governments and the USSR would act as a bridgehead for wider Soviet influence: from fishing agreement to port access to military base (Fry 1991).

Pressure group agitation from the Right sought to make the government take the threat of Soviet influence more seriously. In 1986 retired Major-General Butler declared that he expected there to be a permanent Soviet naval presence in the island Pacific by the turn of the century (Smith 1986). The Fiji coup of 1987 and political unrest in Vanuatu in 1988 created a perception of political instability in the islands, which outside powers may exploit. As late as 1989, the expatriate Australian editor of *The National Interest*, Owen Harries, was ringing the alarm bells at a conference in Melbourne (Harries 1989). On the other hand, some critics believed the government were exaggerating the threat. They argued that the island Pacific was on the way to nowhere and that the USSR had no interests that would justify the cost of attempting to establish some kind of military facility (Fry 1991).

Throughout the 'second cold war', Australia increased its level of aid to Pacific island states as a form of compensation for the islands' rejection of deals with the USSR, and as a way of increasing Australian influence. Australia sought to play the leading political role in the South Pacific forum, keeping the islands 'on side' in the cold war while taking certain island grievances, such as French nuclear testing, to the wider international community. Some island states willingly played along with the idea that they were strategic prizes in order to extract larger amounts of aid from Australia. But others, in particular Vanuatu, were more attracted to ideas of non-alignment and resisted Australia's calls to exclude any contact with the USSR and its allies (such as Cuba and Libya). In 1987 the Australian Foreign Minister, Bill Hayden, rushed from Parliament to make an emergency visit to New Zealand to discuss the apparent crisis in Vanuatu. On his return he found himself so constrained by the need for secrecy that he was unable to explain to a sceptical press what the trip was all about. Both the Foreign Minister and the Minister for Defence discussed the theoretical possibility of Australian military intervention in the island Pacific, and set out what seemed to be loose criteria of intervention (Smith 1992b).

The end of the cold war did away with debates on the strategic significance of the island Pacific and made the policy of strategic denial redundant. There was no longer a serious candidate for 'hostile external power'. Pacific island states had to face their economic marginality to the world economy and to Australia, and a loss of leverage in relations with Australia, now that the idea of a Soviet threat could not be deployed to gain additional aid. Australia's military security is not challenged from this region.

3   South-East Asia

South-East Asia consists of three geographical areas:

- part of continental Asia south of China (Burma, Thailand, and Indo-China)
- the Malay peninsula, where Thailand meets with Malaysia and with Singapore off the southern tip
- archipelagos to Australia's north, which encompass Indonesia, part of Malaysia, Brunei, East Timor, Papua New Guinea, and the Philippines.

Because of its proximity to Australia, South-East Asia has figured prominently in Australian perceptions of possible military threat. Indonesia is Australia's populous northern neighbour, with over 200 million citizens — the fourth largest population in the world. Defence reports since the mid-1980s have focused on the possibility of threats 'from or through' the 'archipelago to our north'. This formulation suggests an ambivalent attitude towards South-East Asia. On the one hand, a threat from the 'archipelago to our north' is code for an Indonesian threat; on the other hand, the idea of a threat through the archipelago implies an Indonesia subdued by an external power (as it once was by Japan and, before them, the Dutch) and suggests Indonesia as a potential Australian ally against such a threat. Both images of Indonesia, as potential threat and potential ally, can find some reference points in Australia's relations with Indonesia since the 1940s.

Whereas the Chifley Labor government of 1945–49 gave support to Indonesian independence, the Menzies Liberal governments after

1949 expressed a growing sense of apprehension about Indonesian foreign policy. This apprehension was partly based on imperial nostalgia, but it was also fuelled by the continuing claim of Indonesia to West New Guinea — that section of the Dutch East Indies that the Dutch had refused to transfer to the new government. The experience of the Second World War led Australian governments to the view that New Guinea was vital to Australia's military security. Indonesia's claims were based on the developing international practice of decolonisation, whereby the post-colonial state inherited the boundaries set by the colonial power. Indonesian parliamentary democracy declined in the 1950s, and in the early 1960s, charismatic nationalist President Sukarno led Indonesia into a militant policy of 'confrontation' with the Dutch over their control of West New Guinea. The USA agreed to support the Indonesian position in the early 1960s, against the wishes of Australia, and West Irian became the Indonesian province of Irian Jaya.

Indonesian foreign policy between 1963 and 1965 then took a more radical turn as President Sukarno sought to 'confront' the British over their proposal to create the state of Malaysia out of its territories in South-East Asia. Australian troops were sent to support the British in 1965. Australia ordered F-111 bombers from the USA in response to the perceived threat from Indonesia. But long before the bombers were delivered to Australia, the perception of Indonesian threat had dissipated. An abortive left-wing coup in Indonesia in 1965 was followed by a counter-coup and the killing of half a million Indonesians by the military and Islamic militants. General Suharto emerged as the new President. Indonesia shifted to a low-key anti-communist, but formally non-aligned, foreign policy, dropping its campaign against the formation of Malaysia.

These dramatic events within Indonesia and the reorientation of its foreign policy from the mid-1960s transformed the perception of Indonesia from one of potential threat to one of potential ally in the cold war. Indonesia became the leader of ASEAN, which by developing economic and political cooperation among the non-communist states of South-East Asia, became the main regional counterweight to the communist states in Indo-China after the withdrawal of the USA from the Vietnam War. With the end of the cold war, ASEAN has developed further as a regional organisation, with Vietnam applying to join, and has extended its agenda to a broad consideration of security issues with the formation of the ASEAN Regional Forum.

Australia's defence review of the mid-1980s articulated the ambivalent attitude to Indonesia in Australia that had developed over the fifty-year relationship. The report stated that:

> In defence terms, Indonesia is our most important neighbour. The Indonesian archipelago forms a protective barrier to Australia's northern approaches. We have a common interest in regional stability, free from interference by potentially hostile external powers. At the same time we must recognise that the archipelago to our north is the area from or through which a military threat to Australia could most easily be posed.
>
> (Dibb 1985, p. 4)

In the diplomatic presentation of its defence policy in the last decade, Australia has stressed that it does not see Indonesia as a threat and that the Indonesian government has welcomed the developments in Australian military policy, which have involved a greater emphasis on 'defending the north'. On the other hand, the methodology used by defence planners seems to presuppose a threat from Indonesia. Official defence thinking claims to focus on the capabilities of states in the region, and the possibility of attack with little warning is seen as arising from those who have the capability to make such an attack. This approach is hailed for its objectivity, as it does not seek to identify any state as a potential aggressor, and it is supported by the argument that, as intentions can change, it is better for the military to remain focused on capabilities. The argument, however, is disingenuous. New Zealand, which like Indonesia has some capability of attacking Australia without warning, is not seen as a potential threat because Australia simply assumes no intention to threaten Australia would ever develop. Despite its ostensible emphasis on the capabilities of other states, Australian defence policy remains principally concerned with Indonesia because of anxiety over possible Indonesian intentions in the future.

Is there a prospect of an Indonesian military threat to Australia? Australia and Indonesia have no historical or territorial disputes of the kind that have led to military conflict elsewhere. Indonesia's external military concerns are to its north, not to its south, and are focused in particular on China.

The major foreign-policy issue that has created a degree of tension in the Australian–Indonesian relationship since the mid-1970s has been the Indonesian annexation of East Timor with extensive loss of life, and continuing conflict and repression. Australian governments, both Liberal and Labor, have recognised Indonesian sovereignty, but this remains a passionately debated position within Australia, and it is resisted strongly by many East Timorese. Recognising Indonesian sovereignty was done in the name of 'good relations' with Indonesia, but ironically the issue has remained the major factor adversely affecting those relations. If Australian diplomacy were to champion East Timorese self-determination, would this have a cost in terms of a deteriorating security relationship with the Indonesian government?

A second issue that may affect the Indonesian–Australian relationship relates to the border between the Indonesian province of Irian Jaya and Papua New Guinea. Indonesian control of Irian Jaya has been opposed by the Free Papua Movement — or Organisasi Papua Merdeka (OPM) — and in the 1980s, Indonesian military actions against the OPM crossed from time to time into Papua New Guinea. In 1984 about 10 000 Irian Jayans crossed the border to Papua New Guinea and sought refugee status. In 1996, 3000 still remained, in villages supervised by the UN High Commissioner for Refugees. Within Papua New Guinea there is a degree of sympathy for the OPM on ethnic and tribal lines. Indonesia's transmigration programs in Irian Jaya may continue to fuel the ethnic conflict and generate support for the OPM. The Papua New Guinean government objected strongly to Indonesian violations of sovereignty, and in 1987 signed a Joint Declaration of Principles with Australia. This was effectively a security treaty, as Australia promised to consult with Papua New Guinea in the event of external armed attack 'for the purpose of each government deciding what measures should be taken, jointly or separately, in relation to that attack' (Evans & Grant 1995, p. 182).

If the 1987 treaty reflected a concern that Indonesia was a potential threat, the government's actions in 1995 reaffirmed the image of Indonesia as a potential ally. In December 1995 the Australian Prime Minister announced, without public or parliamentary debate, that Australia was about to sign a security treaty with Indonesia, which the Foreign Minister said was similar to the treaty with Papua New Guinea and the Five Power Defence arrangements involving Malaysia and Singapore. Article 2, as published in the press, stated that 'The parties undertake to consult each other in the case of adverse challenges to either party or to their common security interests, and if appropriate, consider measures which might be taken either individually or jointly and in accordance with the processes of each party'

(*Australian*, 19 December 1995). The Australian Prime Minister and Foreign Minister declared that the treaty only referred to external security challenges, not to 'internal' issues such as East Timor and Irian Jaya. Criticism focused on the authoritarian nature of the Suharto regime in Indonesia and the prospect that the treaty would nevertheless embroil Australia in repressive Indonesian policies as a result of the closer defence cooperation that would follow.

4   North-East Asia

In shorthand, North-East Asia comprises three Chinas, two Koreas, and Japan. The 'three Chinas' are Taiwan (with a population of 22 million), Hong Kong (with a population of 6 million, returning to the control of the PRC in 1997, and the PRC itself (with a population of 1200 million). The division of Korea dates back to the end of the Second World War and was maintained in the 1950s after an attempt by the communist North to overrun the anti-communist South was repelled by United States troops.

North-East Asia was the source of the only military attack on Australia — by Japan in the period 1942–45. It was also the focus of Australian threat perceptions in the 1950s and 1960s, when fear of China dominated Australia's international outlook. In striking contrast to this history of fear, in the last decade North-East Asia has been identified as the region of greatest promise in Australia's trading relationships. Economist Ross Garnaut's 1989 report *The North East Asian Ascendancy* set out the case for North-East Asia becoming the principal growth area for Australian exports — a prediction confirmed by the mid-1990s. In 1995, five of Australia's top eight export markets were Japan, South Korea, Taiwan, Hong Kong, and China, accounting for 44 per cent of Australia's exports. Are there still military security concerns in Australia about a region that is now so integral to Australia's economic prosperity?

Japan

Japan was Australia's wartime enemy, and it was also the enemy of China, whose eastern provinces Japan had invaded in the 1930s. After the Second World War, Japan's military potential was constrained by an United States-designed constitution, which included the 'pacifist' clause:

> Aspiring sincerely to an international peace based on justice and order, the Japanese people forever renounce war as a sovereign right of the

nation and the threat or use of force as means of settling international disputes. In order to accomplish the aim of the preceding paragraph, land, sea, and air forces, as well as other war potential, will never be maintained. The right of belligerency of the state will not be recognized.

Japan entered a security treaty with the USA and sublimated its militarism into economic growth under Pax Americana. Japan's 'economic miracle' of the 1950s and 1960s saw it emerge as one of the world's leading economic powers and Australia's major trading partner, as coal and newly discovered iron ore were exported to the Japanese steel industry.

Fear of Japan lingered in Australia longer, apparently, than in the USA, and concern about a possible rearmed Japan was one of Australia's motives for signing the ANZUS treaty in 1951. From time to time over the fifty years since the war, politicians and pundits have speculated about the possible revival of Japanese militarism. This anxiety has partly been fuelled by the growing size and sophistication of the Japanese 'self-defence' forces, which although kept within a budget of approximately 1 per cent of GDP and proportionally far lower than any other industrialised state, have grown commensurately with the Japanese economy. Japan's military growth has not, however, caused widespread concern in South-East Asia or further afield, as Japan has not sought the kind of aircraft-carrier-based power-projection capability that the USA, Britain, France, and the former USSR have maintained, and Japan has not sought to acquire nuclear weapons. There is now also a major constraint on the prospect of a reversion to militarism: Japan's economic strength since the Second World War relies on selling to world markets, and sellers usually need to be polite to customers.

In the last decade there have been some signs of growing strategic cooperation between Australia and Japan, with Australia offering to train Japanese peacekeepers. Writer Humphrey McQueen (1991) argued that there is the basis for a much more substantial strategic partnership between the two states.

China, Taiwan, and Hong Kong

After the victory of the Chinese communists over the nationalists in 1949, the nationalist rump fled to Taiwan and was protected from absorption into China by United States military power. Taiwan has since experienced substantial economic growth and become a significant export market for Australia. The nationalists in Taiwan maintained the fiction that they were the true government of all of China — a 'one China' position that the Beijing government has found more palatable than the movement by new political groupings in Taiwan in recent years for independence. Beijing has campaigned earnestly for decades to prevent diplomatic recognition of Taiwan as anything other than a province of China. Taiwan may present an unwelcome picture of capitalist prosperity to the mainland, and has also developed some democratic reforms, but it also acts as an investment conduit into China. This, combined with the probability of United States support for Taiwan against any Chinese military pressure, appears to reduce the likelihood of a Taiwanese–Chinese military clash.

The situation in Hong Kong is more problematic, and at the time of writing China was at the point of reasserting its sovereignty over a territory leased to the British for 100 years. The main question is whether, in the interest of maintaining the considerable economic benefits to China that Hong Kong offers, the Chinese government will allow a degree of economic and political freedom in the territory that will enable it to continue in its role.

China itself has moved through three phases in Australia's perception of military threat: the first phase was influenced by the communist scare of the 1950s and 1960s; in the 1970s and 1980s, it was seen as a cold war quasi-ally against the USSR; and now, since the end of the cold war, it is often viewed with a new foreboding, particularly after the Tienanmen massacre in Beijing in 1991. China experienced sustained economic growth in the 1980s and 1990s, leading to prognoses that it would emerge as the most significant global economic power in the twenty-first century. On the one hand, the capacity for military modernisation would come with economic growth; on the other, a failure in China's economic growth may lead to massive discontent and civil conflict within China. Either way, China would pose significant military security challenges to the region in the next century.

North and South Korea

The two Koreas were a 'hot war' zone in the cold war of the 1950s. Australians fought and died in the Koran War. North and South remain ideological and political adversaries in the 1990s. The South Korean economy, one of the 'Asian tigers', grew rapidly and consistently over several decades and became Australia's second largest export market in the 1990s. Authoritarian politics yielded to a degree of democratic practice. The North Korean economy has stagnated over the same period, and politically it has remained a tight dictatorship. The economic strength of the South has enabled it to develop a hi-tech military force, in response to which North Korea has threatened to acquire nuclear weapons.

North-East Asia and Australian military security

Australia's military security is likely to be affected only indirectly by the foreign and defence policies of North-East Asian states. Three scenarios outline some of the possible crises in this region in future years:

1   Problems could emerge in relation to the clash of interests between China and several states in South-East Asia over the Spratly Islands and accompanying Exclusive Economic Zones in the South China Sea. Australia is not a party to this multi-sided dispute. The main implication for Australia would be the stimulus to military spending by all the parties, creating a more heavily armed South-East Asian region than might otherwise be the case.

2   The reassertion of Chinese sovereignty over Hong Kong and tension over Taiwan may lead to substantial refugee exoduses.

3   The acquisition of nuclear weapons by North Korea may lead to a chain reaction, first with nuclear weapons being acquired by Japan and then, in response to Japan, by states in South-East Asia. A proliferation of nuclear weapons states would constitute a radical deterioration in Australia's military security environment.

5   Indian Ocean

The Indian Ocean region looms much larger in the minds of West Australians than in the minds of the rest of the country. The idea of a military threat from the Indian Ocean region has had two brief incarnations in Australia's postwar history.

In the second half of the 1970s Prime Minister Malcolm Fraser warned of the dangers of Soviet naval expansion in the Indian Ocean, and the invasion of Afghanistan was interpreted as a possible strategic move by the USSR to secure a warm-water port in the Indian Ocean (after securing a corridor through Pakistan). In the late 1980s, there was a brief moment of Australian media focus on the growth of the Indian navy and the 'Indian threat'.

Australian defence policy in the 1980s adopted a 'two-ocean navy' strategy, with a major expansion of naval facilities in Western Australia to supplement the traditional naval bases on the east coast. But this move had little to do with the Indian Ocean region and much more to do with the shortening of naval time to the north of Australia and South-East Asia.

In 1995, Australia hosted a meeting of Indian Ocean states to discuss the prospect of greater economic cooperation and, at Australia's insistence, to consider possibilities of a military security dialogue. The states that might be considered part of the Indian Ocean region consist of those on the west coast of Africa, much of the Middle East, and the states of South Asia, including India, Pakistan, and Bangladesh — a total of more than forty states. While the prospects of economic co-operation gained a hearing, and some initiatives were proposed, the idea of a regional military security dialogue fell on deaf ears. The main reason was clear: the part of the region closest to Australia — South Asia — is preoccupied with its own sub-regional conflicts (conflicts between India and Pakistan over Kashmir, and between India and China over their border and over China's support for Pakistan in its dispute with India).

CONCLUSION

This overview of five regions confirms the assessment of defence adviser Paul Dibb in the mid-1980s that 'Australia faces no identifiable military threat'. What is more, the events of the last decade confirm Dibb's 1986 statement that 'there is every prospect that these favourable security circumstances will continue'. What can Australian foreign policy and defence policy do to maintain this favourable environment?

Antarctica and Oceania, although relatively close to Australia, contain no significant military capabilities. In Antarctica, it may well be that the problems of ecological fragility — relating to scientific programs, to tourism impacts, and particularly to global warming — will generate non-military 'security' issues for Australia that are far more significant than any prospect of Antarctica being the source of future military threat. Rising sea levels may claim more Australian territory than any attacker ever could. In Oceania, without a hostile external power being interested in the islands, there is no prospect of military threat being exercised through these territories. This is not to say that international politics in the island Pacific is likely to be straightforward, or that there are no issues of conflict among the islands, or between the island states and Australia. Some of these conflicts have been generated by Australian-based economic organisations — by mining companies in Bougainville and other parts of Papua New Guinea, for example. Australian military equipment has been used in the war between the Papua New Guinean government and the Bougainville secessionists. For some Pacific islanders, Australia may be a source of threat. In the mid-1990s, Australian military forces were still being trained in intervention strategies and tactics in island environments. Is there a case for maintaining a military intervention option, or are post-cold war international relations in the islands strictly a domain for diplomacy?

Australia's historical experience of threat from, and fear of, North-East Asia may be a poor guide to the kind of military security issues that may affect Australia. The prospect of Japan, China, or India embarking on large-scale territorial acquisition seems remote, partly because of the deterrent effect of the nuclear weapons held, or able to be acquired, by each of them and by the USA and Russia, but also because of the increasing irrelevance of territory to economic power in a globalised economy. Territorial acquisition is no longer 'worth it' in cost–benefit terms. North-East Asia and the Indian Ocean, despite containing several great powers, are regions from which the prospect of direct military threat to Australia is very low because of their distance from Australia, the preoccupation of those great powers with their more immediate regions, and the irrelevance of Australia to their strategic concerns. Again, this is not to deny that there are future uncertainties in these regions, but the strategic implications for Australia of such uncertainties are indirect. If Australia can influence events to maintain 'favourable security circumstances', it will be by Australian diplomacy rather than military defence policy. For Australia to develop defence capabilities to counter a possible invasion or attack by an aggressive, armed North-East Asian power would be to make the mistake of preparing to 'fight the last war'.

In South-East Asia, Australia also faces no current military threat, but proximity to a populous and increasingly powerful neighbour, and the memory of the confrontation policies of the early 1960s, has led to periodic speculation about the re-emergence of an 'Indonesian threat' at some time in the future. It is in the context of the Australian–Indonesian relations that Australia's military forces have a possible role. Australian military force is usually seen as a 'deterrent' to an Indonesian military attack on Australia or Papua New Guinea, and then as a defence if deterrence fails. But military force can have a two-edged effect on international relations. The pursuit of deterrence is counter-productive if it generates reciprocal deterrence on the part of one's neighbour. It may become the impetus for an insidious, if low-key, arms race. A military posture based on deterrence needs to be balanced by a diplomacy of 'reassurance' to prevent a cycle of competition and mistrust from developing.

There are few substantive reasons for supposing a possible Indonesian military threat to Australia. Australia's recent pursuit of 'security as part of the region' is a framework that places foreign policy at the centre of Australia's relations with its immediate neighbour, and the development of a regional security dialogue in the ASEAN Regional Forum may provide a setting in which Australia and Indonesia can avoid becoming a military security problem for each other. The signing of a security treaty with Indonesia in December 1995 was hailed by the then Australian Prime Minister as a 'declaration of trust'. Paul Keating declared that 'We now live in one of the most peaceful parts of the world — and have the machinery in place to keep it that way (*Australian*, 19 December 1995). This would appear to be a major step in the diplomacy of 'reassurance' at a state-to-state level. But the lack of debate within Australia means that the treaty has little grounding in the Australian electorate, which undermines its capacity to build the trust that the government has declared on Australia's behalf.

8 Economic Security

INTRODUCTION: GLOBALISATION

The extent to which any government can provide economic security for its citizens in the 1990s and into the twenty-first century will be largely determined by its capacity to come to grips with the forces of globalisation.

The term 'globalisation' refers to the economic, technological, and cultural processes of global change that have escaped the sovereign control of nation-states. These processes have advanced across national boundaries with astonishing speed and remarkable ease since the 1970s. They include the development of communications technology and media coverage, integrated and deregulated capital markets, the international division of labour, unprecedented levels of economic interdependence, the development of modern instruments of intelligence and surveillance, and the world-wide transmission of ideas and images (Macmillan & Linklater 1995, p. 4).

The development that is of most relevance to this discussion is that which allows private business to shift capital, technology, production, and to a lesser extent, labour around the world to wherever profit can be maximised. In other words, manufacturing centres can be moved to locations in which the cost of land and labour is cheapest, taxation rates are lowest, interest rates are highest, exchange rates are most favourable, and environmental and health regulations are least intrusive. This is possible because the obstacles to the free movement of money and goods around the world have been increasingly removed (through financial deregulation and trade liberalisation) since the early 1970s.

Globalisation has made it increasingly difficult for individual governments to set their own economic policies independently of these extraordinary changes to the nature of the world economy. National economic sovereignty, or the capacity of governments to set their own economic goals and priorities, has been steadily eroding since the early 1970s. The 'disciplines' of world markets impose themselves as irresistible forces on the minds of policy elites, who increasingly recognise the demands of global markets as fundamental constraints upon policy-making. This is true for all governments, no matter how economically powerful they may be.

Globalisation has also changed the very nature of commercial relations between nation-states. Nation-states have traditionally regarded the acquisition of territory as the principal means of increasing national wealth. In recent years, however, it has become apparent that additional territory does not necessarily help states to compete in the international system. In the 1970s, governments began to realise that the wealth of a society is determined by its share of the world market:

> For the conflict is over shares of wealth, and these days wealth is derived not from land, or even from heavy industries that are built up on land. Wealth is derived from the value-added in goods and services sold on world markets. So it is world market shares that are the name of the game for national governments. And as Hong Kong or Singapore or little Taiwan have shown, you do not need more land to be a winner in this new game.
>
> (Strange 1991, pp. 30–1)

Technological change has driven the internationalisation of production. Increasing costs and the need to maintain economies of scale are forcing governments, particularly those with small populations, such as Australia, to look beyond national markets in pursuit of wealth generation. Competing for world-market shares is, therefore, the only way states can generate sufficient wealth to satisfy the demands on their resources and thus keep their societies together. This competition is particularly intense when the world economy is not in a growth phase.

A state requires a scientific and industrial strategy if it is to maintain economic growth and its market shares. Instead of competing for territory and control over people and resources, the advanced industrial states have increasingly begun to compete *on behalf of their producers* for shares of the world market (Strange 1986, p. 21). As was discussed earlier, this relationship is primarily determined by states' dependence on revenue generated by the private sector.

According to Susan Strange, there are three conditions that states must meet before they can play the global game. First, technology has to be mobile and accessible to new players. Second, national markets have to be linked together into one global market by cheaper transport and the faster, cheaper communication of information and ideas. Third, and most importantly for Australia, barriers to the movement of capital have to be removed so that investment is not limited by the supply of local savings. Having met these conditions, new players are then free to compete, however unequally, for market shares of the world economy (Strange 1991, pp. 30–3).

Lester Thurow concurs with Strange's assessment of the changing criteria for 'success' in the modern world economy. He argues that:

> the world [has] shifted from classical comparative advantage, where economic activity depends upon natural resource endowments and factor proportions (capital–labour ratios) — two variables governments have little capacity to alter — to man-made comparative advantage where brainpower industries are central to success. In micro-electronics, bio-technology, the new materials industries, telecommunications, civil aircraft manufacturing, machine tools and computers plus software, government investment is important.
>
> (Thurow 1993)

The forms of international trade have changed. As Hugh Emy argues, 'the idea of national territorially-defined states trading with one another within a conceptually distinct international economy is outmoded' (Emy 1993, p. 168). Intra-industry or intrafirm trade dominates the manufacturing sector of the world economy. Over 40 per cent of all trade is now comprised of intrafirm transactions, which are centrally managed interchanges within transnational corporations. The idea of national and sovereign states trading with each other as discrete economic units is becoming an anachronism. The significance of national boundaries has diminished, as has the power of elected governments to determine economic policy within their own borders. According to one report, 'the world now boasts 37 000 transnational companies, which control about a third of all private-sector assets and enjoy worldwide sales of about $7.5 trillion', a figure just under the USA's GDP in 1993 (*Australian*, 2 August 1994).

These assessments challenge neo-liberal economic theories, which have come to occupy a dominant place in the thinking of Australian governments. Neo-liberalism seeks to minimise the state's role in the economic life of a nation, giving full reign to national and international market forces and ensuring that Australia's domestic policy settings are seen as internationally competitive, especially in terms of efficiency and profit opportunity. For example, in *Australia and the North East Asian Ascendancy*, economist and policy adviser Ross Garnaut presents the neo-liberal case for drawing the state back from protection of the national economy, forcing the economy to internationalise and so gain the benefits of East Asia's dynamic economic growth.

The emergence of intrafirm trade has changed the relationship between governments and private industry. Intrafirm trade runs counter to the theory of comparative advantage, 'which advises nations to specialise in products where factor endowments provide a comparative cost advantage'. The mobility of capital and technology, and the extent to which firms now trade with each other and within themselves, means that 'governments in virtually all industrial societies now take an active interest in trying to facilitate links between their own domestic firms — including offshoots of multinationals — and the global networks' in the strategic industries. From this perspective, governments can no longer remain at arms length from business, as neo-liberal economic theory demands (Emy 1993, p. 173).

The modern trading environment has been increasingly characterised by a struggle for dominance in a number of key strategic industries — micro-electronics, biotechnology, telecommunications — which are crucial to both the viability of modern industrial economies and material living standards in these societies. Success in these industries has depended on reliability, productivity, quality, and generally, the capacity to add value to the commodities to be sold. These industries tend not to be labour intensive, but emerge from the direct application of specialised knowledge gained from technological innovation. States and firms that sustain their international competitiveness in these industries recognise the importance of consumer-driven production, the need to stay abreast of technological developments, and the value of a highly educated workforce. Success and wealth stems from the maximisation of market shares in key industries.

The globalisation of the world economy has seen the spread of high-tech manufacturing industries to many developing countries, and the relocation of transnational manufacturing centres to low-wage, high-repression areas, where access to cheap labour and raw materials is easier. Trading conditions in the 1990s appear to diverge significantly from the assumptions that underpin the neo-liberal analysis of how markets work. Neo-mercantilist theory, which stresses the maximisation of national wealth, also seems to offer an inadequate explanation of contemporary trade realities. A more accurate description of current practices may be 'corporate mercantilism', characterised by 'managed commercial interactions within and among huge corporate groupings, and regular state intervention in the three major Northern blocs (East Asia/Japan, NAFTA, EU) to subsidise and protect domestically-based international corporations and financial institutions' (Chomsky 1993, p. 95).

It is no longer simply a case of states producing national products for national markets. Success no longer depends on a single nation's comparative advantage. There appears to be an important strategic or proactive role for governments in creating conditions that will consolidate a competitive advantage for 'their' firms. In Australia's case, these firms may be foreign-based transnationals, given that two of Australia's three biggest exporters are actually Japanese firms (Mitsui and Mitsubishi).

One consequence of these developments is the steady erosion of the rules which have underpinned multilateral trade in the postwar era. These rules, constituted in forums such as the General Agreement on Tariffs and Trade (GATT), were designed in the immediate postwar period to prevent nations from using trade protectionism and economic nationalism to seek an economic advantage over their rivals. They were designed to discourage bilateral trade deals and provide smaller players, such as Australia, with a chance to influence and shape the global trading environment. It was widely felt, particularly in the USA, that protectionism had been one of the major triggers of the Great Depression of the 1930s. By the 1980s, however, protectionism and neo-mercantilism — in the form of subsidies and non-tariff barriers, such as voluntary restraint arrangements — were again on the rise in Europe and North America. This development has been a concern to primary commodity exporters, such as Australia, which have a greater interest in free trade in primary commodity sectors. However, if the purpose of structural adjustment in Australia is to substantially broaden the country's trade profile and export base, governments will need to form foreign economic policies that take account of the enormous changes in the nature of manufacturing, production, and trade in the globalised world economy.

AUSTRALIA IN THE WORLD ECONOMY: THE HISTORICAL LEGACY

The prospects for Australia's economic security are largely conditioned by the nation's economic history and the patterns of its economic development, especially its colonial inheritance. Brian Head summarises the typical features of what he calls 'dominion capitalism':

> European colonisation was established in semi-temperate lands, sparsely populated by native peoples who were forcibly displaced by large-scale European immigration. These contemporary societies are broadly similar to Western Europe in regard to their high urbanisation and life expectancy, their sectoral distribution of economic activity (primary/secondary/tertiary), and systems and rates of wage labour. However, most of their exports are generated in the primary sector, albeit using advanced technology; industrialisation has been largely confined to processing of primary products and/or import substitution sheltered by protective measures, catering mainly for the domestic market. Transnational firms based in the US, Japan and Western Europe dominate mineral and manufacturing industries; British economic domination was displaced eventually by the United States and at the secondary level by trade and investment links with Japan and the EEC
>
> (Head 1983, p. 6).

Britain's colonial bequest to Australia was to establish it as an exporter of primary products. This has been Australia's economic orientation ever since and seems likely to remain the dominant trade 'reality' for the foreseeable future, as developing a wider base in the export of manufactures and services is a slow and difficult process. As far as world trade is concerned, post-colonial Australia has remained primarily an agricultural and mineral exporter. In recent years commodities have represented over 80 per cent of Australia's total exports — a trade profile more typical of developing economies. This pattern of economic development was established in the early days of British dependency, and has had an enduring effect upon both Australia's internal and external policies. As Thomas Millar argues, 'success in growing wheat and wool, exchanged in Britain for the products of British industry, delayed Australia's own industrial development and its emancipation from dependence on British markets, capital, technology, even economic controls, and foreign policy' (Millar 1978, p. 12). An over-reliance on commodity exports and the absence of a strong export-orientated manufacturing sector are the most significant legacies of colonial capitalism in Australia.

In Australia, for much of the twentieth century, the state played a crucial role in driving the process of industrialisation by building transport and communications infrastructure, attracting labour through publicly assisted migration programs, and borrowing the necessary British capital to finance economic development. Just as importantly, the state also intervened to shield Australia from the rigours of international markets by adopting a wide range of protective measures, including tariffs, quotas, restrictions on foreign investment, and labour regulations. Protectionist policies — which fulfilled the promise of high wages — together with the state's increasing welfare provisions (pensions, unemployment benefits) guaranteed minimum wages and conditions — concessions to democratic participation and to a formalised system of industrial conciliation and arbitration. This represented the high water mark of state intervention in the nation's social and economic life. By the late 1960s Australia had one of the most protected economies in the OECD.

This approach to economic development enabled an infant manufacturing sector to be born. But this manufacturing industry was inward looking, focusing on the Australian market, producing a high cost product, and unable to participate significantly in the postwar expansion of international trade. As the globalisation of production and finance in the world economy got under way in the 1970s, Australia failed to position itself early to take advantage of these important structural changes. Australia's traditional dependence on agriculture and mining exports continued, but in several areas these sectors also faced challenges from protectionist blocs, in particular the EU. Despite its general success in exporting primary products, Australia has acquired a persistent 'current account deficit' on the balance of payments, with receipts from exports unable to match outlays on imports and debt-servicing, creating a continuing need for capital inflow (and overseas debt) to 'balance the books'. Both Labor and Liberal governments have recognised this as Australia's most significant economic problem. Dependence on primary-product exports and on capital inflows have in turn created a heightened vulnerability to fluctuations in world economic activity.

AUSTRALIA'S POLICY RESPONSE

The response of the Australian government to globalisation has been to accelerate Australia's integration into the world economy. From the early 1980s there has been a conscious attempt to remodel Australia's political economy along neo-liberal lines by reapplying the economic principles originally conceived by liberal theorists such as Adam Smith and David Ricardo. At the same time, trade diplomacy has been pursued in order to foster liberalisation on a global and regional basis. According to Gareth Evans, 'it is gradually coming to be appreciated that breaking down such remaining barriers as stand in the way of economic interdependence is a matter of intelligent self-interest; that it is against all experience, as well as all theory, to yearn after autarky in a world increasingly interconnected' (Evans & Grant 1995, p. 12). This explanation expresses the two assumptions that have guided Australian trade policy over the last decade: that the world is slowly but steadily accepting the arguments for free trade and adopting free trade policies, and that there are, indeed, no valid arguments against free trade.

Reducing protection

Since the early 1980s, the idea of exposing Australia to the rigours of international markets has been justified on the grounds that the economy needed to be substantially 'restructured' or 'structurally adjusted' if it was to become 'internationally competitive'. Australia, it was argued, could no longer afford the luxury of introspection, regulation, and protection. It had become internationally uncompetitive — particularly in labour and manufacturing — and faced a bleak economic future unless major structural economic changes were made. The 'discipline' of the international markets would be needed in order to modernise the Australian economy and drag the country into a more secure future. The 1960s and 1970s are characterised as a period of neglect, when instead of developing its manufacturing and services sectors, Australia relied on its comparative advantage in minerals and agricultural production to sustain its relatively high living standards. The economic policies of Labor governments from 1983 were a reaction to the perception that Australia had fallen behind its major trading rivals and partners.

Former Labor governments, supported by the Opposition, adopted a neo-liberal agenda in the belief that this would rapidly modernise and internationalise the Australian economy. They cut tariffs, the major mechanism of economic protection, without waiting for international reciprocation. Other measures involved financial deregulation and 'microeconomic reform' — floating the Australian dollar, de-regulating interest rates and the financial sector, allowing the entry of sixteen foreign banks, and privatising government-owned enterprises (banks, airlines, insurance). Greater efficiencies, increased exports, improved services, and a considerably reduced role for government were the goals of these reforms. The guiding philosophy of policy advisers such as Ross Garnaut was that Australia should only develop industries (and therefore trade) in goods and services, where it had 'natural' comparative advantage. Governments should not intervene in the economy to artificially build industry sectors or 'pick winners', because they would inevitably fail or create inefficiencies that would place a drain on more competitive sectors of the economy. Only market forces can, in effect, make these choices. Industries that could not compete internationally on a 'level playing field' (that is, without government support) had no right to burden the efficient sectors of the economy. This was said to be the only policy option for a modernising economy in a globalised world economy.

Despite sectoral opposition within business communities, tariffs have been reduced across the board to unprecedentedly low levels, most significantly in the textiles, clothing and footwear, and motor vehicle industries. The result has been greater efficiencies and higher quality products, with the prospect of continuing investment in successfully restructured areas, but significant job losses.

Promoting free trade internationally

While the state stepped back interfering in areas within its sovereignty, it has not taken such a passive approach internationally, and Australian governments in the 1980s and 1990s set out to promote and support reforms of the international trading regime. A liberal global trading regime, free from state subsidies and protection, was seen as the best way of securing Australia's economic well-being in the future.

In the mid-1980s Australia embarked on an ambitious task of multilateral economic diplomacy, attempting to link together global agricultural exporters in the battle against agricultural protection, particularly protection of the EEC. As part of the Cairns Group of 'fair' agricultural traders, Australia argued at GATT meetings (multilateral trade negotiations now conducted under the auspices of the newly formed World Trade Organisation) for the phasing out of agricultural subsidies. In theory, Australian rural exports would substantially increase if they were free to compete on an equal (unsubsidised) footing with European and United States producers.

The Australian government's recognition of the economic importance of the East Asian region was belated, but was nevertheless symbolised eventually in 1989 by the formation of the APEC group, comprising fifteen regional economies that are committed to 'an open, multilateral trading system in the interest of Asia-Pacific and all other economies' and to reducing 'barriers to trade in goods and services among participants in a manner consistent with GATT principles'. Australia played a significant role in the formation of APEC.

APEC has already passed through a number of phases in its short life. In 1989 it was ostensibly sold to the world as an OECD-type organisation, designed to promote both regional trade liberalisation and closer economic links between Australia and the fast-growing East Asian economies. However, the unstated agenda of the initiative, was that, if the 'Uruguay Round' of multilateral trade negotiations collapsed, APEC possessed the embryonic structure of a future trading bloc under Japanese leadership. Hence the USA was initially excluded from membership of the organisation. Once the USA insisted on joining APEC, and was accepted for the inaugural ministerial meeting in September 1989, APEC's secondary purpose was effectively off the agenda.

Since then, APEC has evolved, at least in theory, into an open-ended free trade zone. It has a growing bureaucracy, to be based in Singapore, and a widening trade and economic agenda, which is regularly discussed at the heads of government/economy level. From an Australian perspective, APEC is seen as a 'GATT overlay', which can work in harmony with the multilateral trade talks by promoting global liberalisation, and which will eventually provide a broad overarching framework encompassing both an expanding NAFTA and the ASEAN Free Trade Area (AFTA). Whether it is possible to have 'free trade' within a regional organisation that does not discriminate against non-members, however, is a question that is yet to be answered. All APEC members agree that the organisation should not become a trading bloc like the EU, with a common currency and tariff walls to keep out the rest of the world. But the idea of an 'open regionalism' that does not create a preferential trade grouping is difficult to imagine. The alternative — which would give the Europeans a 'free ride' into the Asia Pacific — is equally unattractive to some of the major players, particularly the USA. By 1994 there was already a growing division within APEC between 'developed members' (Australia and the USA, for example), who were pushing for an early free trade agreement that would liberalise trade and investment, and 'developing members' (such as ASEAN), who were less enthusiastic about lowering trade barriers, and preferred to emphasise development and technical cooperation within the group. Although it refuses to explicitly commit itself, the Australian government still appears to favour preferential trade liberalisation (discriminating against non-APEC members) over a non-discriminatory agreement. Attempts by Australia to consolidate APEC by renaming it an 'economic community' in time for the first leaders summit in Seattle were met with fierce resistance by 'developing' members of the organisation, particularly Malaysia.

Despite its initial objectives, APEC is said to play an important role for Australia by locking Japan and the USA into the one free-trading structure. However, fears shared by Australia and a number of developing economies in the region that they could be caught in the crossfire of a trade war between Japan and the USA have only eased slightly, and there is little evidence that APEC yet provides a 'circuit-breaking' structure for resolving ongoing United States–Japanese trade tensions. Similarly, APEC apparently has no role to play in resolving the dispute over intellectual property rights between the USA and China, which threatens to bring about another 'trade war'. The first leaders summit (without Malaysia's prime minister, Dr Mahathir, in attendance) was held in Seattle in November 1993 and was largely the achievement of the Australian Prime Minister at the time, Paul Keating. Keating believed the organisation required the involvement of heads of government, or 'leaders' in the case of Taiwan and Hong Kong, if it were to be effective in promoting trade liberalisation.

The second APEC leaders summit, held at Bogor in Indonesia in November 1994, was another milestone in Australian trade diplomacy. If the reaction of the Australian media is any indication, the second APEC leaders summit was an extraordinary success. The eighteen members of APEC, with a combined population of 2.2 billion, a gross output of US$12 trillion, and generating 41 per cent of world trade, declared their commitment to 'free and open trade and investment at the latest by 2020'. This would be accomplished by a two-stage process, with industrialised countries reaching the target of free trade by 2010 and developing countries being given an additional ten years to complete the task. It was up to each member to decide which category they fell into, and deciding the means by which each economy would reach the goal was put off until the third leaders summit in Osaka, twelve months later. The declaration signed at Bogor was a non-discriminatory agreement without internal preferential arrangements. 'Free trade' was left undefined.

The Prime Minister specified the benefits that would flow to Australia from the second APEC leaders summit. As a result of the Bogor declaration, Australia would receive A$7 billion in additional income each year and a boost in its economic output by 3.8 per cent — this is said to be twice the size of the benefit that is expected to accrue from the signing of the Uruguay Round of GATT in December 1993 (if these benefits were to eventuate). An extra 70 000 Australian jobs would also be created as the barriers to free trade are lowered and Australia's exports to the region increase. Just as importantly, at Bogor, Australia secured a permanent seat at perhaps the world's most important diplomatic table. Australian prime ministers could now look forward to annual summits with the leaders of the world's most dynamic economies.

The third leaders summit held in Osaka in November 1995 did little to advance the process beyond reaffirming the targets set at Bogor. No universally agreed approach for reaching the two free trade goals was decided upon, although the fact that no backsliding had occurred, particularly in agricultural sectors, was considered to be important. In addition, Japan and China made significant 'down payments' on trade liberalisation, reinforcing their commitment to the Bogor target figures. Again, the Prime Minister claimed that Australia's national income would be boosted by 6.8 per cent, or $40 billion a year, by 2010 and that 500 000 additional jobs would result from free trade in the region (*Australian*, 20 November 1995). Malaysia, the least enthusiastic member, stressed that it regarded the agreements reached at Bogor and Osaka as voluntary and non-binding. The difficult work of establishing explicit timetables for achieving free trade among the group was put off until the 1996 leaders summit, to be held in the Philippines.

EVALUATING AUSTRALIA'S POLICY RESPONSES

Australia's national economic policies and regional and multilateral trade diplomacy have been oriented around a consistent set of neo-liberal assumptions about the path to economic security in the contemporary world economy. When in Opposition, the Liberal-led coalition supported the broad direction of the ALP's efforts, leading to a new bipartisanship on free trade and an open economy, which replaced the old bipartisanship on the need for industrial protection.

It is tempting to suggest that this view reflects the government's commitment to liberal internationalism. The liberal view that free trade and commerce can unite the world by generating greater wealth for all has been around since the industrial revolution. It may well have been this idea that recently inspired the states of Western Europe to seek economic and political integration beyond the traditional nation-state formula. But, for Australia, neo-liberalism is a recognition that small and medium powers are influenced by world market forces and global economic trends that they can do very little to control. Resistance, in the form of protectionism, while popular, has proved somewhat futile and costly, encouraging inefficiency and uncompetitiveness, although this view is contested. In fact, the revival of neo-liberalism can be seen as a response both to assaults by the international financial markets *and* to the inept interventionist behaviour of the Australian state in the postwar period (Bell 1993).

With regard to Australian public policy, the desire for a more open economy and a multilateral free-trading system is the prevailing orthodoxy of the moment. It is built around a belief that protectionism has failed and that market forces operating through foreign competition will improve Australia's economic performance. According to neo-liberal assumptions, Australia's comparative advantage in agricultural and mineral production can only be maximised by promoting trade liberalisation in the hope that this will prevent the global economy fragmenting into closed trading blocs.

How can the cost and benefits of this approach be evaluated? And what are the alternatives? The positive side to these changes has been a boost in Australia's exports, particularly in financial and educational services, although also in some manufacturing sectors. Unfortunately Australia's appetite for imports has also increased, thereby nullifying any significant improvement in the economy's current account deficit and balance of payments problem. The negative aspects of these changes suggest there is a continuing need to explore alternatives, although the current bipartisanship is resistant to the idea that alternatives are needed. Several lines of criticism are discussed in the rest of this chapter (see also Higgott, Leaver, & Ravenhill 1993).

Loss of economic sovereignty

Recent financial deregulation appears to have significantly eroded the capacity of Australians to control their economic destiny. Power over economic decisions is increasingly placed in the hands of currency traders, speculators, credit ratings agencies, stockbrokers, lenders, and foreign (transnational) investors. As Australia is incapable of generating sufficient endogenous wealth to finance its economic development (that is, to maintain the standard of living most of its citizens expect), governments need to provide domestic economic conditions that will attract foreign investors to the country. In a world in which capital markets are globally linked and money can be electronically transferred around the world in microseconds, states must consider their comparative 'hospitality' to foreign capital — that is, they must offer the most attractive investment climates for the investment of relatively scarce supplies of money. This gives the foreign-investment community enormous influence over the course of a nation's economic development, and constitutes a significant diminution in a country's economic sovereignty.

The volume of foreign exchange trading (the buying and selling of money) in the major financial centres of the world, estimated at US$640 billion per day, has come to dwarf international trade: foreign exchange trading is at least thirty-two times greater. The brokers on Wall Street and in Tokyo, the clients of the 'screen jockeys' in Westpac foreign exchange, and the auditors from credit ratings agencies such as Moody's and Standard & Poors can now pass daily judgements on the management of the Australian economy and signal to the world's financial community the comparative profit opportunities to be taken advantage of through investment in the country. Inappropriate interventionist policies on the part of government can be quickly deterred or penalised with a threatened or actual reduction in the nation's credit rating, or a 'run' on the nation's currency. The needs of the international markets must be anticipated at all times.

Finance markets are amorphous entities and, like their psychology, are difficult to define. The bond markets, for example, were recently described as 'a loose confederation of wealthy Americans, bankers, financiers, money managers, rich foreigners, executives of life insurance companies, presidents of universities and non-profit foundations, retirees and people who once kept their money in passbook savings accounts and now buy shares in mutual funds' (*Australian*, 25 June

1994). It is clear that the larger banks and financial institutions, insurance companies, bond holders, funds managers, brokers, and speculators are the dominant players. But predicting the behaviour of money markets on a daily, if not hourly, basis — which is, after all, what financial speculation is all about — is a perilous activity and only for the stout-hearted. The one and only thing that can be said with any confidence about money markets is that they will act in their own interests — as they perceive them. That is to say, they will act to maximise their own wealth (hence the pressure on the Reserve Bank of Australia at different times to increase local interest rates).

Another government response to both the need for foreign investment and the growing debt crisis (currently standing at A$163 billion or 42 per cent of gross national product) is to 'sell off the farm'. Consequently, foreign ownership of Australian companies has reached 35 per cent, making Australia, next to Canada, the most foreign-owned country in the OECD. Foreign equity ownership across all sectors of Australian industry is already at 32 per cent and is estimated by the Economic, Planning and Advisory Committee (EPAC) to rise to 50 per cent. The policies of deregulation introduced in 1983 — the privatisation of state-owned enterprises, a relaxation of foreign-investment regulations, the abolition of exchange-rate controls, foreign-bank entry, and the deregulation of interest rates — have greatly accelerated this process. The attitude of the Foreign Investment Review Board (FIRB) — a somewhat secretive branch of Treasury charged with protecting 'the national interest' in questions of foreign takeovers of Australian businesses — is instructive. FIRB support for foreign acquisitions is now virtually routine, with estimates putting the approval rating as high as 98 per cent. In 1992–93, foreign investment in Australia jumped by a third to A$24 billion (*Australian Financial Review*, 3 March 1994). Despite these figures, any suggestions of restricting foreign investment, favouring endogenous capital, or reviewing foreign-investment guidelines are met with swift opposition from the bureaucracy and the financial press. Reregulating the financial sector in the community's interest is viewed in the business press as 'coercion' and a 'threat' (*Australian Financial Review*, 19 July 1994). Economic nationalism is not considered an appropriate topic for public discussion in the 1990s.

These decisions were not reluctantly made in the face of the overwhelming pressures of the world economy. They were taken with enthusiasm in the belief that these reforms would rapidly modernise the nation's economy, which had become bloated and inefficient behind the barriers of protection. The Australian economy is now, more than ever, exposed to changes in world commodity prices, movements of foreign capital, and alterations in exchange rates. It may be no exaggeration to suggest that, in the 1990s, Australia is a hostage of the world economy. In the era of the 'borderless economic world', Australia cannot afford to lose the confidence of international markets. Loss of sovereignty was the price to be paid for international efficiency and a faith in free trade.

Domestic and international consequences

Neo-liberal theory discusses the benefits that are to flow to 'Australia' from deregulation without considering the distributional consequences. The reforms have not yet secured the sharp reduction in unemployment and underemployment which are supposed to eventuate in the 'long run' but have not materialised in the 'here and now'. The most notable social consequences of policies colloquially described as 'economic rationalism' include growing disparities in wealth distribution and income, and increases in family and child poverty (Stilwell 1993). Opinion polls suggest that the overwhelming majority of Australians opposed free trade (*Sunday Herald-Sun*, 28 February 1993; *Age*, 23 April 1993, 9 June 1994). These social consequences may partly explain the Labor government's election loss in 1996, although the Liberal-led coalition seemed to promise essentially the same recipe.

We must ask if, internationally, a policy of free trade is consistent with the government's advocacy of environmental protection, international humanitarianism, and the protection of labour rights? If environmental regulations are declared illegal under WTO guidelines, is the Australian government prepared to sacrifice Australia's native forests in the interests of regional trade liberalisation? If freedom of association, minimum wage and safety standards, and independent trade unions threaten East Asia's 'comparative advantage' in low wage costs, will the Australia government curtail its free trade advocacy? Or would this undermine its support for APEC?

Increasing global protectionism?

The principal problem with Australia's free trade diplomacy is its assumption that the rest of the world also wants to move in the same direction. All Australia apparently needs to do is convince the Europeans and Americans that free trade is also in their interests. In reality, there is little evidence that any of the trading blocs actually want a 'level playing field', or that they are prepared to move significantly towards one. While paying lip service to the idea of free trade, the dominant economic players tend to practise a 'mixture of liberalisation and protection, designed for the interests of dominant domestic forces, and in particular for the TNCs [Transnational Corporations] that are to run the world economy' (Chomsky 1993, pp. 94–5). For wealthy industrial societies, free trade is often non-reciprocal.

Unsurprisingly, GATT negotiations were stalled for a number of years on the issue of agricultural subsidies, which were estimated to cost Australia A$3 billion in exports every year. The final signing of the Uruguay Round agreement in December 1993, while offering few, if any, benefits to the Third World, will provide a boost for agricultural and mineral exporters such as Australia. Though optimistic predictions assert that Australia can expect A$5 billion a year in increased exports, 50 000 more jobs, and a A$3.75 billion rise in its annual output as a result of the round's successful conclusion, there is no reason to believe that the agreement signals even the beginning of the end of agricultural subsidies: the agreement only requires subsidies to be cut by up to one-third. The power and influence of European farmers ensured that the Uruguay outcome was substantially less beneficial to so-called 'fair traders' than it might otherwise have been, and that the concessions that required protected traders to roll back their subsidies were, in the end, minimal. (*Australian Financial Review*, 12 and 13 April 1994). In fact, a recent World Bank report has claimed that the Uruguay Round achieved little or no liberalisation in agriculture; 'dirty tariffication' may even have increased levels of protection, resulting in negligible income gains for Australian farmers (*Australian Financial Review*, 7 April 1995). With only limited progress at the most recently concluded round of multilateral trade negotiations, largely the result of fierce resistance from French farmers, Australia may eventually have to rethink its commitment to the multilateral trading system.

The current neo-liberal orthodoxy also fails to acknowledge that some of the inherent problems of resource-based economies have nothing to do with inefficiencies — problems such as deteriorating terms of trade (a decline in the price Australia receives for its exports) and the social impact of reductions in industry protection. As Hugh Emy and Owen Hughes have remarked, 'Australia could easily have the most open economy in the world and still be worse off' (1991, p. 69).

Alternatives for a more interventionist state

Other approaches may need to be considered. A bilateral free-trade agreement with one of the major blocs — North America (NAFTA), Europe (EU), or a Japanese-led block — is another possibility. Perhaps the Cairns Group of free-trading agricultural exporters or the APEC group foreshadows the formation of another trading bloc? There will be increasing pressure upon governments to consider these options if the promised dividends from the Uruguay Round's completion are not realised.

Strategic government planning in industry and trade policy is another alternative. The most 'successful' East Asian economies have had their industrial development coordinated, to a substantial degree, by the state. The postwar economic revivals or 'miracles' that have occurred in these states can be directly attributed to their preference for communitarian forms of capitalism, and to their failure to subscribe to the artificial separation of state and economy, a popular principle with neo-liberals in Anglo-Celtic societies. As Emy explains:

> the strongest industrial economies, those most successful in global competition, were those where the state played a carefully designed strategic role aimed at both fostering the internal competitiveness and productivity of the domestic economy, while assisting its most efficient (and strategic) industries to acquire and consolidate a competitive advantage in the global economy — especially in those areas of manufacturing whose products were increasingly central to the viability of high tech economies and the living standards of their citizens.
>
> (1993, pp. 9–10)

The impressive performance of the East Asian economies has been achieved only 'partly through their pursuit of efficient resource allocation via the unfettered price mechanism (the neo-liberal model), and owed rather more to the role of the state in deliberately guiding or governing the way in which their respective market economies developed'. Cooperation between business and government to gain world market shares in selected industries has been the rule rather than the exception (Emy 1993, p. 193). Trevor Matthews and John Ravenhill, after an extensive study of 'strategic trade policy' in East Asia, conclude that

'State intervention to enhance the benefits that domestic economies gain from participation in global production chains may be decisive in determining the technological trajectories that countries will follow' (1993). Australia could therefore consider more sophisticated and selective forms of state intervention in the economy, where industry and labour can be quarantined from the vicissitudes of the international markets, without compromising the need for industry efficiency and export competitiveness. Government and industry may need to work more closely together to assist firms to win and consolidate shares of the global market. In other words, following the East Asian example, Australia might need an interventionist, and necessarily risky, industry policy in place of the high-risk strategy of reliance on markets to deliver an optimal economic outcome.

CONCLUSION

In the absence of any perceived military threat, and in response to a decline in material prosperity, the focus of Australia's policy-makers has belatedly switched to the creation of wealth. Integration into the world economy is now said to be the only way Australia can increase its export trade and attract foreign investment. Australia did not take advantage of the postwar expansion of international trade and failed to move away from its over-reliance on commodity exports. It is only now responding to the impact of the globalisation of the world economy and beginning to face up to some of the new realities of international trade. The choice of direction, particularly the policy response to the changing way in which 'economic success' is achieved in international trade and finance, will largely determine Australia's economic future and social composition.

9   Environment

INTRODUCTION

In the post-cold war years of the 1990s, there is a general perception that an unprecedented and unsustainable level of global environmental degradation exists. It is now widely accepted that, to deal with this situation, states need to cooperate on policy formulation and implementation. With the shift away from a solely military conceptualisation of security, there now appears to be a willingness on the part of most states to discuss what has become known as 'environmental security'. States, through international political organisations like the UN, have begun coordinating global strategies to combat a range of environmental problems. Among the more prominent of these problems is global warming.

Even though the interstate system is presently characterised by a high level of interdependence, this does not necessarily mean that effective multilateral policy on the environment will be forthcoming. States will, as they have in the past, continue to operate on the basis of their interests. Multilateral policy on the environment will have to steer a course among competing sets of national interests. There is a real danger, therefore, that international political organisations will remain constrained in their efforts to formulate environmental policy initiatives.

This chapter examines the Australian government's role in the establishment of the United Nations Framework Convention for Climate Change (FCCC) from a foreign policy perspective. It contrasts the government's strong initial support for the FCCC in 1992 against its actions to undermine the provisions of the FCCC during the Berlin conference in 1995. Several issues are pertinent to this discussion. First, we will look at global warming within the context of the 'global environmental crisis', demonstrating that global warming is an

181

issue requiring urgent attention. Existing multilateral policies on climate change, it will be shown, are inadequate. Second, the chapter argues that the concept of sustainable development lacks sufficient acknowledgment of the social, political, and economic issues involved in 'growth first' prescriptions for environmental protection. This particular policy orientation — one that the Australian government has embraced fully — is likely to create new environmental problems and exacerbate existing ones. Third, the Australian government's retreat from its earlier commitment to reducing greenhouse gas emissions underscores the potential weakness of multilateral policy as currently framed. It also highlights the flaws in the government's assertions that the notion of 'good international citizenship' informs its foreign-policy orientation.

CRISIS? WHAT CRISIS?

While the planet is perhaps yet to reach *an* environmental crisis, a multitude of environmental crises presently exist. It is important to note that these individual cases are part of a system-wide problem. As Jim MacNeill argues:

> Globally we are in the process of skinning the planet alive. Forty years ago, Ethiopia had a 30-per-cent forest cover; twelve years ago, it was down to 4 per cent; today, it may be 1 per cent. Seventy-five years ago, India's forests covered over half the country; today, they are down to 14 per cent. In 1961, Thailand's forests covered 53 per cent of the country; twenty-five years later, in 1986, they covered only 29 per cent — and were going fast. In the tropics today, 10 trees are being cut for every one planted. In Africa, the ratio is 29 to 1.
>
> (1989–90, p. 6)

The removal of forests — which act as important 'sinks' for greenhouse gases — has the potential to hasten global warming. Global warming in turn has the potential to change the environmental conditions within which all nation-states — indeed, all planetary life — function.

Is there sufficient evidence to argue that global warming, created by the emission of greenhouse gases (particularly carbon dioxide and methane), will inevitably lead to a rise in the sea level and increase the rate of desertification? The public demand for governments to respond quickly to global warming (Kempton 1991; Brechin & Kempton

1994) is an issue of concern for some in the scientific and policy-making communities (Singer 1992; Norse 1994, p. 147). These groups have argued that governments need to wait for more concrete evidence on global warming. Peter Gleick takes up this issue:

> An alternative approach is to wait for more research and more detailed regional information on the environmental and economic impacts of climate changes before taking preventative actions. This approach has two serious flaws. First, the complexity of modelling climatic behavior means that the necessary research will be slow and difficult. Unless actions begin soon to reduce the emissions of carbon dioxide and other gases, the earth will be irreversibly committed to substantial warming. Second, any international agreement to prevent major climatic changes may be complicated by a desire of certain actors (alliances, nations, sub-national groups, corporations) to capitalize on perceived regional advantage.
>
> (1989, p. 310)

The interrelationship between different elements in the environment means that small changes may affect the total environment in ways not yet fully understood (Hurrell 1994, p. 148). In this context, scientific knowledge and forecasting would seem to be fundamentally important. To wait for conclusive evidence, however, will only delay the process of modifying human activity. Jeremy Leggett and Paul Hohnen argue that valuable time has already been lost in multilateral negotiations on climate change (1992, p. 76). Allowing this process to continue may create a situation in which it could be too late to intervene (Mische 1989, p. 393). At that stage, no level of policy coordination would succeed in remedying the situation.

It has been known for some time that climatic tolerances are remarkably fine. Barbara Ward and René Dubos noted in the early 1970s that 'It may require only a very small percentage of change in the planet's balance of energy to modify average temperatures by 2°C. Downwards, this is another ice age, upwards, a return to an ice-free age. In either case, the effects are global and catastrophic (1972, p. 266). Changes to the earth's climate that appear to be minor have the potential to create significant difficulties for many communities. Rising sea levels present low-lying island states with a particular problem. Many of these countries could vanish beneath the ocean. But low-lying island nation-states would not be the only victims of rising sea levels. Over one-third of the world's population inhabit the coastal regions. In Australia that percentage is higher: three-quarters of Australians live within 50 kilometres of the coast. Coastal regions are also significant sites of agricultural and industrial production. Higher sea levels mean that tidal surges and what are now 'manageable' storm patterns will result in greater levels of destruction. Some estimate that 30 million people in Bangladesh alone could become environmental refugees as a result of rising sea levels (Patterson 1993, p. 185).

There is an obvious need to develop policies that will, at a minimum, stabilise current rates of global warming. The effectiveness of these policies will depend on the political decisions of states and on the level of international cooperation. Therein lies a significant problem. In a world comprised of competing sovereign states, will it be possible to develop a multilateral policy that protects the global environment and to which all states will conform?

THE MULTILATERAL PROBLEM

One of the strongest claims that the environmental movement makes is that 'it will be extraordinarily difficult to arrive at ecologically sound solutions to the world's environmental problems within the context of a global system comprised of militarized, privatized sovereign states' (Seager 1993, p. 144). Joe Camilleri and Jim Falk claim that the present interstate system is in the process of significant change: 'The theory of a world partitioned into territorial domains under the superintendence of national states has enjoyed considerable currency over the last century, but is increasingly confronted by another equally long-standing perspective of a living planet as an integrated whole' (1992, p. 192). Richard Falk argues that, in opposition to the state-centric world, a new form of global politics — social movements that embody an ecological ethic — will emerge (1987, p. 191). Yet in the 1990s it seems that states are, in fact, multiplying. It is unlikely that states will, in the short to medium term at least, be replaced by a new form of 'planetary' politics.

As Andrew Hurrell correctly notes, even if it were possible to do away with states, the 'political and moral tensions that have long been central to international society ... would not go away, but would merely reappear in a different form'. Moreover, it is important not to view 'the inability of human beings to agree on the nature and seriousness of environmental threats, on the meaning of sustainability, and on the principles of global environmental management' as being 'primarily rooted in the state system' (1994, p. 165).

If environmental issues are to be dealt with effectively within the existing state system, it is important that the processes through which policy is formulated and articulated be open to scrutiny. Two issues are important. First, there is the need to develop a more nuanced analysis of the state and its relationship to environmental issues (Conca 1994; Patterson 1995, pp. 221–2). Second, it is necessary to understand the way in which particular interests dominate multilateral forums. The idea put forward by some — for example, James Rosenau (1993) — that ecological interdependence will force the cooperation of recalcitrant states should be dismissed as little more than wishful thinking. Unless vigorously monitored and made accountable, states will not unilaterally act to protect the 'global', as opposed to the 'national', environment.

Australia's former foreign minister, Gareth Evans, talks enthusiastically about the increasing attention given to environmental issues by policy-makers and multilateral institutions (Evans & Grant 1995, p. 11). Evans asserts that the international community, specifically foreign-policy-makers, can develop programs to manage these 'big problems' (Evans & Grant 1995, pp. 11–12). He argues that the appropriate conceptual framework for dealing with environmental problems is sustainable development. Evans argues this on the basis that sustainable development 'rejects the false dichotomy between economic growth and the protection of the environment' (Evans & Grant 1995, p. 164). Because sustainable development is a central theme running through the government's approach to international and domestic environmental policy, it is worth discussing this issue in some detail.

SUSTAINABLE DEVELOPMENT: A WAY FORWARD OR A CONTRADICTION IN TERMS?

Sharon Beder notes that the concept of 'sustainable development' has become central to most states' approaches to environment issues since the UN World Commission on Environment and Development (WCED), chaired by Gro Harlem Brundtland, released its report, *Our Common Future* (Beder 1993, p. xiii). In this document, the commission stated that sustainable development could be seen as 'a process of change in which the exploitation of resources, the direction of investments, the orientation of technological development, and institutional change are all in harmony and enhance both current and future potential to meet human needs and aspirations (WCED 1987, p. 9). A broad statement such as this, which clearly lacks specificity, has been the subject of wide and varied interpretation. As a consequence there remains strong disagreement over the precise meaning of 'sustainable development' (Jacob 1994, p. 241; Visvanathan 1991).

Explicitly rejecting the work of those environmentalists who argue that there are distinct environmental limits to economic growth, the proponents of sustainable development claim that the concept is about the 'growth of limits' not the 'limits of growth' (Martell 1994, pp. 15–40). The WCED noted that 'sustainable development involves more than growth. It requires a change in the content of growth, to make it less material and energy-intensive and more equitable in its impact' (1987, p. 52). Technological innovation may assist in reducing resource and energy inputs, but the argument that growth should be made more 'equitable' raises broad social, political, and economic issues. The distribution of wealth and resources among developed and developing countries is a useful point of entry into this political economy issue.

Our Common Future, although acknowledging that 'A world in which poverty is endemic will always be prone to ecological and other catastrophes' (1987, p. 8), fails to deal with this issue of global political economy in a systematic and critical manner. The following extract from the report underscores the point: 'A necessary but not sufficient condition for the elimination of absolute poverty is a relatively rapid rise in per capita incomes in the Third World. It is therefore essential that the stagnant or declining growth trends of this decade be reversed' (WCED 1987, p. 50). Merle Jacob states that the 'growth first' emphasis of the report was finally accepted, despite the many submissions arguing against this position made to the WCED during the preparatory stage of the report (1994, p. 245). William Graf claims that the report needs to be interpreted as part of a political process, rather than merely accepting it as a technical or environmental document. Graf points out that, like other UN-sponsored reports, *Our Common Future* was 'Written by politicians rather than environmentalists. He argues that 'these reports, memoranda and appeals are significant less for their intrinsic theses and analyses than for the hegemonic ideological projects that they represent and the particularist interests that underlie them' (1992, p. 553). Given these criticisms, it is important to examine the report in its particular political and historical context. The work done by Peter Doran is useful here:

> The World Commission on Environment and Development ... was set up at the height of the Cold War in 1983. Brundtland saw issues of the environment as a way forward that avoided the East–West political gridlock. The Cold War may have ended but it lives on in the text of 'Our Common Future' (1987) which applies a security analogy with origins in that period. Environmental degradation, together with the problems of development such as poverty, were said to be a threat to security. The response offered was a mobilisation of all available means to exterminate the threat, leading to a resource, risk and crisis management approach.
>
> (1993, pp. 63–4)

Doran is correct to view the work of the WCED within a cold war versus post-cold war context. In so doing, it becomes clear that environmental issues have not radically redefined the core issues of international politics. What has occurred is that environmental issues have become incorporated within an essentially status quo international political framework, where they have become problematised as 'security issues'. Patricia Birnie's work alludes to this issue when she observes that 'environmental threats can be equated with other threats to international security, such as military aggression, and can be dealt with by the Security Council under Chapter VII of the Charter' (1995, p. 68). Unlike Ian Rowlands (1991), Daniel Deudney (1990) argues that there is something inherently misguided in conceptualising environmental problems as issues that can be accommodated within the security paradigm. Dominant notions of sustainable development conform to a particular political framework. Is there any benefit in attempting to redefine sustainable development?

Rethinking sustainable development

Simon Dalby challenges the assumptions that underpin the current usage of the term 'sustainable development'. He offers an alternative meaning of sustainable development and, in so doing, explicitly rejects orthodox notions of what he terms 'development': 'Sustainable development has to be converted into meaning the sustainability of ecological fecundity and species diversity rather than the sustainability of the industrial system and the state structures that support it. But then, of course, if it were to take on this meaning it would have long ceased being "development"' (1992, p. 117). Although Dalby offers little guidance on how this transformation can be achieved, the point he makes is pertinent. A challenge to the legitimacy of existing notions of sustainable development is required. A reformulated notion of sustainable development that brings into question present forms of production and existing property rights will, however, find strong opposition among those elites, classes, and groups in developed and developing countries who benefit from unequal exchange relations. In a period of globalisation, where deregulation and privatisation policies are promoted as the means of economic growth, efforts to protect the environment will be resisted by those states eager to maintain their 'share' of the world's wealth (Sanders 1990).

One area that requires particular attention is the link between the WCED's model of sustainable development and the almost total acceptance of economic rationalist principles in some policy-making circles in relation to environmental issues. J. Sinner provides a useful example of this form of reasoning when he states that:

> it can be efficient for a polluting industry to shut down in a country with high environmental standards and relocate to a country with low standards ...
>
> In this situation, both countries would gain from a shift of production from the first country to the second. The shift increases global welfare because the same product would be produced at less total cost given the respective costs of pollution in the two countries.

(1994, p. 179)

Permitting the continuation of 'free market forces' to organise the global economy, as it has in the past, will only result in even greater levels of environmental degradation. The 'growth first' model, which underpins the present concept of sustainable development, is flawed at four levels. First, it fails to recognise that past patterns of economic growth have brought the planet to its current state of ecological degradation. Second, it is based on the optimistic assumption that the global political economy — if the WCED rhetoric about allowing developing countries to reach the economic standards of the developed economies is genuine — will be able to grow at a rate far exceeding previous rates of economic growth. The implicit assumption in this argument is that the current global recession is cyclical, not systemic. It also disregards the economic pressures on Third World states still labouring under the burden of massive debt repayments amassed during the mid-1980s (Buttel & Taylor 1992, pp. 224–5). Third, there is the extremely contentious assumption that the planet can sustain existing, or increased, levels of pollution. Fourth, 'growth first' implies that, in developing countries, economic growth will eventually raise the living standard of the poor. Historical experience shows that the 'trickle down' process tends to produce highly uneven patterns of economic growth within states. Rapid industrialisation in developing countries is more likely to generate large-scale environmental degradation, which will impinge adversely on resources — land, timber, and feed stocks — essential to the material well-being of the poorest in those communities.

The WCED report *Our Common Future* was released in 1987, and its significance cannot be overstated (Conca 1993, p. 311). Most Western states have now come to accept the particular interpretation of sustainable development as articulated in the report. *Our Common Future* also formed the basis of a number of high-level UN conferences on the environment. Perhaps the most comprehensive and widely known of these was the UN Conference on Environment and Development (UNCED) — the 'Earth Summit'. In June 1992, in excess of 170 countries met in Rio de Janerio to discuss global environmental issues.

PROMOTING INTERESTS: AUSTRALIAN FOREIGN POLICY AND GLOBAL WARMING

This section presents a case study that seeks to tease out some of the themes raised in the preceding pages, showing the way in which the Australian government formulated its response to the global environmental problem of climate change. First, it will be shown how the government accepted and then endorsed a particular interpretation of sustainable development drawn from the work of the UN WCED. Second, this section examines Australia's position at the 1992 Earth Summit. At that conference, Australia was party to a multilateral agreement to reduce greenhouse gas emissions. The final part of this section deals with Australia's attempts to 'undermine' targets and timetables discussed at the Earth Summit on the grounds that these goals would carry a domestic economic and political cost.

An important point needs to be made at the outset. A binding multilateral commitment to achieving targets and timetables was not reached at the Rio Summit. As Ted Hanisch correctly claims, 'Commitments that would mean a halt — or indeed reduction — in total global emissions of Greenhouse gases (GHGs) was never on the Agenda ... The most binding commitment proposed but not agreed upon was for the developed countries, individually or jointly, to stabilize emissions by the year 2000 at 1990 levels' (1992, p. 64). The Australian government was, therefore, not under any strict obligation to reduce national greenhouse gas emissions. Moreover, as Ian Lowe has noted, 'the government has explicitly stated that it will only attempt to meet the target if there will be no negative effect on the economy generally or competitiveness in particular' (1994, p. 316). In that sense, the charge that the government 'undermined' or 'retreated' from commitments given at Rio may at first blush appear to be unwarranted. In the following pages, it will become evident that the government was, initially at least, committed to significant reductions in greenhouse gas emissions.

Embracing Brundtlandism: *Our Country Our Future*

In July 1989 the Prime Minister, Bob Hawke, announced the government's first comprehensive statement on the environment. *Our Country Our Future* provided a document that would see government policy on the environment coordinated across ministries. *Our Country Our Future* openly recognises that significant environmental destruction has taken place in Australia as a result of economic development since European settlement: 'Nearly two-thirds of the continent requires treatment for land degradation. Forest cover, just 10 per cent when European settlement began, has been halved. More than 41 million hectares of forest have been destroyed, including 75 per cent of the nation's rainforests' (Hawke 1989, p. 1).

As with *Our Common Future*, the Australian government's *Our Country Our Future* regards economic growth as the cornerstone of sustainable development. In an interesting twist, the 'no growth' argument is challenged on the grounds that such a policy would harm developing countries: 'Poverty itself is a major source of ... environmental problems' (Hawke 1989, p. 3). There are clear echoes of *Our Common Future*. 'Those who are poor and hungry', the WCED report noted, 'will often destroy their immediate environment in order to survive' (1987, p. 28). The links between the WCED report and the Hawke government's *Our Country Our Future* are developed in the following quotation from Hawke:

Fortunately, the challenge in front of us is not a stark choice between preserving the environment and economic growth. As the World Commission on Environment and Development (the Brundtland Report) has pointed out, we have the ability to make development ecologically sustainable. The task is to ensure that we meet the needs of the present without compromising the ability of future generations to meet their needs.

Ecologically sustainable development means economic growth that does not jeopardise the future productive base. Renewable resources are managed so that they are not permanently depleted. In some cases the use of particular technologies or processes may be so damaging that they should be banned.

Only rarely will it be necessary to take such pre-emptive action. In most cases it will be sufficient to temper the way in which projects proceed or technologies are applied to ensure that our future productive base is not impaired.

(Hawke 1989, p. 4)

An incremental and technocratic approach to environmental policy threads its way through *Our Country Our Future*. As evident in the WCED report, there is an overwhelming bias towards the preservation of existing economic and social practices. In this document, the Australian government commits itself fully to the 'principle of ecologically sustainable development' (Hawke 1989, p. 4).

Australia at the Earth Summit

The Earth Summit represented the culmination of many years work by the UN to place environmental issues on the international agenda. The momentum for this process started as early as 1972 with the UN Conference on the Human Environment held in Stockholm. In 1987 the WCED released *Our Common Future*. In the same year, the UN conducted a conference that produced the Montreal Protocols on Substances that Deplete the Ozone Layer, calling for the multilateral phasing out of chloro-fluorocarbon (CFC) use. The Montreal Protocols stemmed directly from the Vienna Convention for the Protection of the Ozone Layer of 1985. Global climate change was first addressed within the UN system by the Intergovernmental Panel on Climatic Change (IPCC), established by the United Nations Environment Programme (UNEP) and the World Meteorological Organization (WMO) in the late 1980s.

Two important sets of documents were developed at the Earth Summit. First, the meeting reached agreement on an international program of action to protect the environment. *Agenda 21* was designed to be a practical guide to action that all states could undertake within their nations (United Nations Conference on Environment and Development 1992; Robinson 1993). States would be required to report to the UN Commission on Sustainable Development (UNCSD) and show how domestic policies contributed to the *Agenda 21* framework (Commonwealth of Australia 1995). Second, the gathering at Rio was presented with two conventions that, if states chose to ratify, would oblige states to adopt particular environmental policies. Let us now examine the FCCC in detail.

Framework Convention on Climate Change (FCCC)

The FCCC was motivated by the findings of the IPCC, which in the late 1980s released data claiming that, if current emissions of carbon dioxide continued, by 2025–2050 there would be a doubling of the carbon dioxide levels in the atmosphere. This could, it was argued, result in a significant rise in global temperature — between 1.5°C and 5°C — which could produce a subsequent rise in sea level in the order of 0.3–0.5 metres by 2050, and about 1 metre by 2100. In addition, the report maintained that if this 'business-as-usual' approach were to continue, there would be a rise in the temperature of the surface ocean layer of between 0.2°C and 2.5°C (IPCC 1990). By the time the Earth Summit was convened in 1992, there existed a substantial amount of research material that could be presented to delegates.

With much of the analysis and policy already thrashed out, the Earth Summit was, in fact, a meeting at which states would negotiate the context, meaning, and interpretation of the draft convention. That process, as Sten Nilsson and David Pitt note, was one in which the final framework convention came to mean 'many things to many people' (1994, p. 55). The major source of these differing perceptions was Article 4/2 (a) of the FCCC. In its final and agreed form, this section of the FCCC states that:

> Each of these Parties shall adopt national policies and take corresponding measures on the mitigation of climate change, by limiting its anthropogenic emissions of greenhouse gases and protecting and enhancing its greenhouse gas sinks and reservoirs. These policies and measures will

demonstrate that developed countries are taking the lead in modifying longer-term trends in anthropogenic emissions consistent with the objective of the Convention, recognizing that the return by the end of the present decade to earlier levels of anthropogenic emissions of carbon dioxide ... would contribute to such modification ...

(Commonwealth of Australia [undated], p. 18)

Nowhere in this subparagraph is there established a definable level at which states would agree to set their carbon dioxide emissions. The commitment that states would reduce carbon dioxide emission rates to 1990 levels is, in fact, contained in subparagraph (b) of Article 4/2. However, subparagraph (b) stipulates that it must be read with reference to subparagraph (a), as well as Articles 12 and 7. This means that the relevant subparagraph on gas emissions, Article 4/2 (a), is subject to so many caveats that in the end it becomes a 'most convoluted and opaque' statement (Nilsson & Pitt 1994, p. 25). Even when these caveats are overlooked, subparagraph (b) is still loose enough to enable broad and varied interpretation. Article 4/2 (b) states that signatories are to frame their policies 'with the aim of returning individually or jointly to their 1990 levels' (Nilsson & Pitt 1994, p. 19).

The FCCC does not compel states to act in any manner other than to promote their own narrow self-interest. Indeed, that was precisely how Article 4/2 (a) and (b) was formulated. The USA refused to sign the FCCC if precise targets and timetables for the reduction of carbon dioxide emissions were included. Other states refused to sign unless specific targets and a timetable for action were set. In the end, Britain negotiated a compromise arrangement, which provided sufficient leeway for all states to interpret the FCCC according to their own particular agendas (Nilsson & Pitt 1994, p. 25). That said, it should not be assumed that the FCCC was seen by all present as a meaningless document. Compromise may have been needed in order to enable the most profligate carbon dioxide-emitter to be incorporated into the FCCC, but even with its poorly specified targets, the FCCC was considered by many as an important step in the right direction. This was the Australian government's position (*Age*, 12 June 1992).

Australia, the good international citizen

Given the strength of comments made by the then Minister for the Environment, Ros Kelly, on her return from Rio, the expectation was that Australia would not be one of those states to renege on these limited targets. Portraying Australia as a 'vigorous and respected "middle" power, a coalition builder that gets results', Kelly noted that:

> Australia and many other countries worked and negotiated for stronger outcomes on climate change, biodiversity and some of the items in Agenda 21 — but in the end getting global agreement acknowledging global environmental problems, and setting out on the road to action is one [sic] of the biggest and most important international agreements since the establishment of the UN itself. That it is a UN outcome is, of course, also important.
>
> Australia supports the [FCCC's] call for developed nations to stabilise greenhouse emissions, and considers the reporting mechanism for both developed and developing countries a major opportunity for continued pressure to bring about real action.
>
> (1992, pp. 2, 3)

In addition to these public statements, a more permanent commitment had previously been given by Australia's Foreign Minister, Gareth Evans, on the issue of multilateral environmental policy. That commitment took the form of Australia's inclusion in the Hague Declaration of 1989. Elizabeth Russell has written that 'the addition of Australia's signature to the Declaration of the Hague was a significant indication that the environment was being integrated into foreign policy objectives and responsibilities' (1994, p. 7). Noting that urgent action was required on the issue of global climate change, the first principle of the Hague Declaration was to develop 'within the framework of the United Nations, [a] new institutional authority ... which, in the context of the preservation of the earth's atmosphere, shall be responsible for combating any further global warming of the atmosphere and shall involve *such decision-making procedures as may be effective even if, on occasion, unanimous agreement has not been achieved*' (Hague Declaration 1989, emphasis added).

Gareth Evans has previously acknowledged that the Hague Declaration specifically recognises that 'industrialised countries have special obligations to assist developing countries which will be negatively affected by changes in the atmosphere' (1990, p. 114). Australia's response to international environmental issues is 'an integral part of the broader foreign policy interest we have in being — and being seen to be — a good international citizen' (Evans 1990, p. 113).

As David Goldsworthy (1994) notes, the 'idea' of good international citizenship is firmly fixed within the national interest. According to Evans, rising sea levels would produce environmental refugees from the low-lying states in the South Pacific, 'who would look mainly to Australia for resettlement' (1990, p. 113). Note here that Evans perceives environmental problems such as climate change as 'threats' to Australia's security. To combat these possible 'threats', Evans advocates that Australia needs to 'promote universal adherence to those [multilateral environment] conventions already negotiated', and argues further that 'we need to develop new framework conventions on the protection of the atmosphere and climate change' (Evans 1994d, p. 117).

On the one hand, it is important to recognise that no binding commitment was developed at Rio. On the other hand, it is clear that the Australian government made a firm commitment to take action on the issue of climate change, both through its public statements — its stated disappointment that stronger measures had not been agreed to at Rio — and by its signing of an international instrument like the Hague Declaration. The government was serious about multilateral policy on climate change, and hence its actions in 1994 and 1995 do mark a significant reversal of its previous position.

Backpedalling to Berlin

By late 1994 the Australian position on carbon dioxide emissions changed dramatically. In replacing Ros Kelly, Senator John Faulkner now began the process of moving away from the original commitment that was so enthusiastically given at Rio. Speaking to the 'Greenhouse 94 Conference' in New Zealand, Faulkner argued that 'if we continue with our current level of greenhouse response, Australia will be seven per cent above our 1990 levels of greenhouse gas emissions by the year 2000. It doesn't take a scientist to deduce that, the way we're going, we won't be able to meet either the national or international greenhouse gas reduction targets we have agreed to' (Faulkner 1994, p. 5).

The problem for the government was that, having signed and then ratified the FCCC, it now found itself in a position in which it would have to defend its all-too-obvious retreat before the Conference of the Parties to the FCCC (COP-1) in Berlin, scheduled for late March 1995. Immediately prior to leaving for Berlin, Faulkner launched the Commonwealth government's additional greenhouse response package: Greenhouse 21C. The emphasis of

Greenhouse 21C was to make it clear that, while the government had a particular responsibility for carbon dioxide emission control, it could not fulfil these targets without the cooperation of industry. Faulkner stated that: 'The Government welcomes the commitment industry has shown over the past few months. They have shown a willingness to make substantial and measurable reductions in emissions. I believe the community has an expectation that industry will play its role. And the Government expects industry to deliver' (Faulkner 1995, p. 1).

A criticism that can be levelled at the government was that it had the responsibility to establish policy guidelines and regulations that would fulfil the commitments it gave at the Earth Summit in 1992. In the three years after Rio, the government did little to regulate carbon dioxide emissions. The release of the National Greenhouse Response Strategy in December 1992, coinciding with Australia's ratification of the FCCC, contained no firm policy or strategies on how Australia would meet the FCCC target on greenhouse gas emissions.

During the latter part of 1994, the government appeared to be embracing the 'user pays' principle for those industries and activities that emitted carbon dioxide gases as a way of encouraging polluters to install new technologies and systems to reduce their rates of emissions. This became known as the 'carbon tax', to be set at the rate of A$1.25 per tonne of carbon dioxide. Export industries were exempted from the pollution levy. Only those industries that consumed fossil fuels within Australia would be subject to the carbon tax. Australian business groups rejected the proposed tax and reinforced this position by threatening to cancel investment projects where carbon dioxide emissions would incur penalties.

After considerable negotiation, the government dropped the proposed carbon tax in mid-February 1995. Incensed by the government's obvious lack of willpower in committing itself to the FCCC agreement, Australian environmental groups argued for an end to government subsidies for the forestry and mining sectors. If the government was not going to regulate carbon dioxide-emitters, the environmentalists argued, then government subsidies to carbon dioxide-generating activities should cease. Business too remained dissatisfied —not with its obvious success in having the carbon tax defeated, but with the government's general inability to establish an integrated and non-contradictory environmental policy around which businesses could plan their investment strategies. As the executive director of the Electricity Supply Association of Australia had earlier observed, 'The Federal Government doesn't have an all-embracing understanding of what is going on, or a central strategy. It has simultaneously in place a policy on micro-economic reform to drive the price of electricity to the cheapest in the world, and a greenhouse abatement proposal likely to lead to ever-increasing taxes on electricity' (*Australian Financial Review*, 19 January 1995).

With the COP-1 in Berlin now only a matter of weeks away, the government's strategy for compliance with the FCCC lay in tatters. The key plank in that strategy had been the carbon tax. Business interests were partly responsible for the demise of the carbon tax; however, John Faulkner also faced strong opposition from Cabinet colleagues. The economic ministers rejected the carbon tax. The Treasurer, Ralph Willis, publicly opposed the levy and went even further, indicating that the government was likely to consider ways that it could evade its perceived FCCC obligations. Willis cautioned:

> We are concerned with ensuring that Australia does everything in its power to try to live up to its obligations to the convention, under which it accepted that it should try to keep its emissions to 1990 levels by the year 2000. Obviously there are some let-out clauses and those are not unimportant clauses. They have to be taken into account when considering whether we need absolutely to tie ourselves to achieving the 1990 emission levels and, if not, on what basis we would decide that something less than 1990 emission levels are appropriate for us.
>
> (House of Representatives, *Debates*, 7 February 1995, p. 582)

Without Cabinet support for his carbon tax initiative, John Faulkner had to prepare a position for Berlin that would reduce criticism of Australia's retreat from its earlier FCCC commitment. Australia's UN Ambassador for the Environment, Penelope Wensley, had the task of explaining the government's position at the opening of the Berlin conference. Wensley argued that the 'weight of evidence on the pace of global warming' meant that stabilisation policies were too weak (*Sydney Morning Herald*, 31 March 1995). The government's strategy was now to argue that the FCCC had become too compromised and needed strengthening. Taken at face value, this appears to have been a direct contradiction of Australia's desire to break its FCCC commitment and a return to its position at Rio.

In reality, this tactic was used to reopen the debates on targets and timetables. This gave the Australian government an opportunity to build a coalition with other states who also wanted to dilute the FCCC commitment. Japan, the USA, Canada, Australia, and New Zealand — a group known by the acronym JUSCANZ — campaigned to have the FCCC commitments renegotiated. For its part, the Australian delegation argued that, while it accepted the proposition that the developed world should bear the principal burden for reducing carbon dioxide emissions, the developing countries — which under the FCCC were not required to reduce emissions of greenhouse gases until the year 2000 — should now be subject to greater controls.

On the eve of the Berlin conference, the Australian Bureau of Industry Economics released a document maintaining that Australia's contribution to lowering greenhouse emissions under the FCCC was excessive. Arguing for more 'burden sharing', the report noted that if developing countries reduced emissions, then Australia would be able to dramatically lessen its FCCC commitment (*Australian*, 28 March 1995). Despite Faulkner's protests that Australia 'was not "sleazing or sliming" away from the question of targets', the Australian policy faced considerable hostility at Berlin (*Australian Financial Review*, 5 April 1995). After exhaustive negotiations, the other members of the JUSCANZ group finally reached agreement. Australia, however, remained adamant that emission targets were detrimental to its national interests. A compromise position was brokered. Unlike the original FCCC negotiations at the 1992 Earth Summit, in which the USA had proved to be the difficult party, one of the most enthusiastic supporters of the 1992 FCCC — Australia — now sought to scupper the original obligations and contain any development of the targets it had earlier ratified.

Faulkner's plan to include some developing countries — Brazil, Korea, Malaysia, and China — in the pool of states that would limit carbon dioxide emissions was not a strategy to reduce overall levels of global greenhouse gases, but to offset Australia's share of carbon dioxide emissions reductions. Faulkner failed in this attempt, but the Berlin conference did defer the process of setting strict targets and timetables. The outcome at Berlin was little more than an agreement to engage in talks about setting goals for the reduction of greenhouse gases — known as the 'Berlin Mandate to Negotiate a Protocol'. The Berlin Mandate explicitly recognises that some states have different economic structures and resource bases, which would need to be taken into consideration when emission levels and compliance dates are to be discussed. Unlike the Australian Coal Association, Faulkner, as environment minister, could not publicly celebrate this aspect of his contribution at Berlin (*Australian*, 10 April 1995). Yet Faulkner professed to be pleased with the results of his efforts (*Canberra Times*, 9 April 1995).

CONCLUSION

There are a number of important observations that can be drawn from this chapter. The Australian government's approach to global environmental issues has been motivated more by the desire for cost displacement than an aspiration to play a genuine multilateral role. To what extent, therefore, can it be argued that the Australian government takes multilateralism seriously? This is a large and complex question. It is clear that in its policy approach to global warming the Australian government has pursued national interests before notions of the 'common good'. This distinctly realist approach cuts directly across the government's pronouncements of 'good international citizenship', which it maintains is a fundamental part of its liberal internationalist foreign-policy approach.

At one level, multilateralism appears to offer states an opportunity to work towards an environmental program that will ensure cooperation among states. This chapter has shown how the Australian government reneged on an international agreement to reduce greenhouse gas emissions. Motivated by trade and economic issues, and an unwillingness to formulate domestic policy that would bring it into confrontation with business, the government undermined an international agreement on the reduction of greenhouse gas emissions. Its strategy to reduce the targets and extend the timetables for compliance with the FCCC represents a particularly cynical use of multilateralism.

Evans's statement that industrialised countries have a particular responsibility to assist developing countries by ensuring that environmental problems do not create adverse conditions for these states stands in contrast to Australia's actions at the Berlin conference, where Senator Faulkner argued that some developing countries should 'share' more of the 'burden' by reducing their greenhouse gas emissions. As this chapter has argued, this strategy was aimed at reducing global levels of greenhouse emissions, which would excuse Australia from meeting its accepted targets. In this context, it is important to note that the government has remained careful not to openly use the non-binding aspects of the 1992 FCCC as a mechanism to release it from its perceived obligations.

A recurring theme throughout the government's foreign policy is the role of Australia as a 'middle-power coalition-builder'. Indeed, the government celebrates this role. In the case of the negotiations on climate change, the government certainly fulfilled the role of a coalition-builder. At the Berlin conference the Australian government constructed a coalition to overturn the provisions of the 1992 FCCC. While this may not be the 'coalition-builder' image with which the government wants its foreign-policy 'style' to be associated, it is one that its actions have made it difficult to avoid.

Future negotiations will almost certainly revisit some of the assertions made at previous conferences, and arguments about the difficulty of setting strict targets and timetables will re-emerge. It will be interesting to see whether the Australian government has a set of workable policies in place by then to control greenhouse gas emissions, or whether it will yet again seek to undermine multilateral environmental programs such as the FCCC.

10 Human Rights

Human rights pose a particular problem for students of international politics. On the one hand, there is the notion that only states can confer rights on members of the nation. On the other hand, the conceptualisation of human rights clearly transcends the jurisdictional boundaries of individual states. The tension is therefore between citizenship of a particular state and a recognition of universality — membership of humanity. Thus human rights sit uneasily between the domestic and international realms. As a descriptive term, 'human rights' appears to be more closely aligned to the universalist principle, promoted by proponents of global community, that individuals should enjoy rights not because of their location within a nation-state but because they are 'human'. Yet the ability to enjoy those rights, both realists and rationalists would argue, is bound up in the political, social, economic, legal, and cultural structures of individual states.

To promote universal human rights as a durable feature of global politics, then, challenges the first principle of the extant interstate system: sovereignty. The imposition upon states of duties and obligations that are derived externally from international covenants and treaties clearly undermines states' ability to determine matters within their borders. This uneasy, perhaps unresolvable, relationship between the state and the system on the issue of human rights is reflected in the Charter of the UN. Although it is the principal legislature with regard to universal human rights, the UN nonetheless has a fundamental tenet of the interstate system enshrined in its charter: the domestic jurisdiction principle.

One of the most immediate human rights issues that now confronts the international community is the refugee crisis. Over the past six years the ranks of the global refugee population have swollen dramatically. Yet unlike the political and economic environment of the 1950s — in which refugees became an accepted part of the postwar immigration policies of countries such as the USA, Canada, Australia, and New Zealand — in the world of the 1990s, refugees find that the opportunity to migrate is restricted. For the vast majority of refugees, entry to their country of choice is denied. To understand why this transformation has taken place and its likely consequences for Australia's foreign policy, we need to examine the political context of the past fifty years. To begin with, let us consider the theoretical perspectives of international relations outlined in Chapter 1.

INTERNATIONAL RELATIONS THEORY AND HUMAN RIGHTS

For realists, human rights are merely one of the many transitory issues of international relations. Legitimate though it may be for states to deploy human rights to maximise advantage in a competitive system, states are not, nor should they be, accountable to any supranational authority. States must remain free to choose to uphold, or choose to ignore, human rights. In that context, therefore, the end of the cold war is of no great significance, except, that is, to those realists who rejoice in defeating a cold war opponent whose abuse of human rights (as defined by liberal-democratic traditions) added to their demise. Where the national interest demands, human rights will be overridden by the *raison d'être* of the state.

Removed from the specific conditions of the cold war *per se*, the issue of human rights represents for realists something that is more an enduring and systemic feature of the interstate order itself. John Vincent argues:

> There is an inescapable tension between human rights and foreign policy. Their constituencies are different. The society of all humankind stands opposed to the club of states, and one of the primary rules of the latter has been to deny membership to the former. Foreign policy, according to these rules, should be conducted among states. It should not involve itself either with the communities endorsed by states, or with the notional global community which reformers, revolutionaries and other trouble-makers have called up to justify their enthusiasms. The society of states should and does concern itself with rights, but they are not the rights of individuals, or even of nations, but of states.

(1986, p. 129)

A system comprised of national units with the ability (realists would argue the sole responsibility) to create enforceable legal codes will, almost certainly, remain the focus of attention for human rights claims. Nevertheless, balancing this realist scepticism is the rationalist perspective, which considers global politics to be shifting away from a rigid state-centric perspective to a more nuanced and multilateral form.

Moving towards a position in which human rights will become an integral part of the interstate system, rationalists maintain that the evolutionary development of international law will eventually produce universalised human rights. David Beetham explores this theme:

> we inhabit two worlds, or two paradigms, simultaneously. One is what Forsythe calls the 'anarchical society' of individual states, the other the 'global governance' of international standard setting; as he shows, foreign policy is conducted within both these worlds. Rosas writes of the transition from the ... system of state sovereignty to a multi-layered system regulated by universal principles of law. Although this second world, to which human rights belong, is still embryonic, its development has been cumulative.
>
> (1995, p. 4)

Thus, rationalists point to an unfolding process through which human rights have, over time, been manoeuvred towards a central place in the governance of the international order. Progressively, states will be brought to the recognition that it is in their interests to adopt human rights principles. International law and multilateral diplomacy are the instruments that will bring peace and security to the anarchical world of competing states by incorporating human rights within the international legal framework. To some extent, the rationalist optimism can be regarded as springing from two factors. First, released from the ideological fetters of cold war politics, states will increasingly be both able and willing to assent to internationally agreed standards of human rights. Second, greater levels of interdependence remove the ability of particular states to evade penalties for non-compliance with internationally agreed human rights directives. For rationalists such as Australia's former foreign minister, Gareth Evans, reforming current international institutions, such as the UN, to reflect these new circumstances is vital (Evans 1993, pp. 3–16). Similarly, as the current UN Secretary-General asserted in 1992:

The foundation-stone of this work [reform of the UN] is and must remain the State. Respect for its fundamental sovereignty and integrity are crucial to any common international progress. The time of absolute and exclusive sovereignty, however, has passed; its theory was never matched by reality. It is the task of leaders of States today to understand this and to find a balance between the needs of good internal governance and the requirements of an ever more interdependent world.

(Boutros-Ghali 1992, p. 9)

At both the national and international levels, then, the rationalist perspective would appear to dominate the policy-making elite's perceptions of how to deal with human rights issues. Yet it is important to recognise that this so-called 'internationalism' is nonetheless firmly located within a system in which the principal solutions are state-based.

Global-community visions of world order are not synonymous with a reformed UN. Cosmopolitans maintain that the irresistible force of 'one worldism' will outgrow the machinations of state-centric power politics, which has so long been the prevailing condition of international relations. Critical social and political movements could thus become important agents of global social change. Joseph Camilleri develops this argument:

Despite wide variations in size, constituency, organisational structure and ideological outlook, these movements have had a cumulative impact on the shaping of national and world public opinion. Human rights are now central to political discourse; they have become an integral part of the diplomatic agenda. As a consequence states are obliged to formulate comprehensive human rights policies even though these are often more honoured in the breach than the observance.

(1989, p. 89)

The obvious retort to the global-community vision is that, while these movements have pushed and coerced states and international institutions to take up issues such as human rights, they lack the power to enforce the goals they seek. The reply is that there is merit enough in the *promotion* of human rights; when engaging in a political process characterised by unequal power relationships, *any* positive outcome or reform is significant. Whether these movements can endure and help transform the global society remains uncertain, for as Robert Walker readily acknowledges, 'their future has to be achieved' (1988, p. 31).

However, there can be little doubt that these movements have influenced many states with their constant vigilance and surveillance of human rights issues.

THE UNITED NATIONS AND HUMAN RIGHTS

Human rights in the UN Charter

Shaped by the experiences of the Second World War, the UN — unlike its predecessor, the League of Nations — 'explicitly listed human rights as a principal concern of the new organization' in its charter (Donnelly 1992, p. 251). Although, as we shall see, there are strong arguments for reading the various Articles of the UN Charter as interdependent, it is useful to focus on the two Articles that deal specifically with human rights. These can be found in Chapter 5 of the Charter, 'International Economic and Social Co-operation':

Article 55
With a view to the creation of conditions of stability and well-being which are necessary for peaceful and friendly relations among nations based on respect for the principle of equal rights and self-determination of peoples, the United Nations shall promote:
a   higher standards of living, full employment, and conditions of economic and social progress and development;
b   solutions of international economic, social, health, and related problems; and international cultural and educational co-operation; and
c   universal respect for, and observance of, human rights and fundamental freedoms for all without distinction as to race, sex, language, or religion.

Article 56
All members pledge themselves to take joint and separate action in co-operation with the Organization for the achievement of the purposes set forth in Article 55.

Although the Charter lists the obligations of its member states to promote human rights (Article 55) and takes action to ensure that human rights are universally achieved (Article 56), the debate over the precise intent of the Charter has raged since its formulation. The debate has largely been between those who consider that the specific, narrow direction given in the Charter should prevail (legal positivists) and those who favour a broader interpretive position (natural law theorists):

> On the one hand Kelsen has argued that the Charter does no more than set out the purposes and functions of the UN in relation to human rights; it imposes no obligations on Member States. On the other hand Lauterpacht argues that the obligation of Member States to promote human rights includes an obligation to protect them. This is because recognition of human rights is a constant and dominant feature of the Charter, and Member States are required to act in accordance with its purposes. The fact that human rights are not specified in the Charter does not, in Lauterpacht's view, derogate from the legal nature of the obligation undertaken.
>
> (O'Neill & Handley 1994, p. 114)

Yet the UN Charter is only one aspect of the UN's promotion of human rights. The UN Commission on Human Rights was established in 1946, and by 1948 the General Assembly of the UN had adopted the Universal Declaration of Human Rights, which was, in fact, not a treaty but a resolution of the General Assembly. Extending the rights articulated in the Universal Declaration, the UN developed the International Human Rights Covenants in 1966. This process, principally aimed at conferring legal status on the signatories to the Universal Declaration, came into force in 1976.

The most successful of the UN directives on human rights has, therefore, been the progressive development of human rights covenants. These instruments have sought to extend and give specificity to those Articles of the Charter that refer to human rights. Nick Poynder details the UN covenants that relate to human rights: 'In a world-wide context, human rights are measured by the standards set out in the International Bill of Human Rights. This document consists of three instruments: the Universal Declaration of Human Rights, the International Covenant on Economic, Social and Cultural Rights, and the International Covenant on Civil and Political Rights [ICCPR]' (1993a, p. 60).

The development of UN covenants dealing with human rights issues suggests that the framers of the Charter, although clearly incorporating human rights, were unable to give full weight to the relevant Articles because many states opposed any binding clauses. As should be expected, the process was one of political compromise. It would be misleading, however, to view the UN as having adopted a human rights agenda as a direct consequence of its members having taken an enlightened position since that time. As Theo van Boven notes, non-government organisations have played a critical role in promoting human rights, often by publicly embarrassing transgressor states (1991, p. 183). In 1961 Amnesty International was formed. Now with a membership in excess of one million, Amnesty continues to highlight human rights abuses through its comprehensive monitoring and reporting systems. Moreover, the pressure to adopt human rights has been driven, in important ways, by the movement for political independence that emerged at the end of the Second World War. The Afro-Asian states, who by the mid-1960s formed the largest voting bloc within the General Assembly, took a particular stance on human rights as a direct consequence of their colonial experiences. Hence the progression of the human rights agenda within the UN has been the result of political pressures coming from differing quarters.

It is important to note that a significant discrepancy exists between the strong monitoring role of the UN and its lack of an unambiguous ability to enforce. 'The strongest "enforcement" power available to any of these bodies', Donnelly notes, ' was (and remains) adopting a critical public resolution or report' (1992, p. 252). Enforcement and implementation of human rights still remain prerogatives of individual nation-states. Thus, securing human rights covenants has not been an easy task. In a very real sense, the problem has been one of extracting human rights from the Charter in which there are no specific directives, but in which there are numerous implied duties and obligations. As with all international political instruments, compliance will ultimately be determined by issues other than particular legal or philosophical interpretations of treaties or agreements. In the conflict and compromise endemic to the interstate system, it is hardly surprising that interests play a major role in delimiting commitment to human rights. The framing of the UN Charter is a case in point, not least because some of the UN's founding members sought to limit the powers of the organisation from its inception. Australia was a leader among those states.

Australia's UN connection: the domestic jurisdiction principle

At the San Francisco conference — where the basic framework for the emerging UN system was developed — the ALP's minister for external affairs, (1941–49) Dr Herbert Vere Evatt, argued strongly for the inclusion of a principle that would disable the UN from intervening in matters that could be argued were 'internal affairs' (Renouf 1983, p. 250). Undoubtedly Evatt was conscious of the emerging pressure for postwar decolonisation, and thus he was likely to have been motivated by a desire to keep Australia's trusteeship of Papua New Guinea, the 'White Australia' policy, and the treatment of Aboriginal Australians away from possible international criticism.

Australia's geographical location forced upon it a particular sensitivity to these sorts of political issues. And so too Australia's wartime experience endowed it with a view that its security interests would be best served by denying a potential aggressor territorial acquisition at its northern approaches. The combination of these interests led to Evatt's determination that the domestic jurisdiction principle — Article 2(7) — be included. Once embedded in the Charter, this principle reinforced the notion of inviolable state sovereignty. In relation to promoting human rights, this principle has been a significant impediment. Gareth Evans correctly observes that:

> The development of the U.N.'s human rights institutions and agenda since 1945 has involved the gradual overriding of initially strict views about non intervention in internal affairs. The initial condemnation of apartheid by the General Assembly in 1952 was an important milestone, as was the creation of the special rapporteur system in 1967 and the introduction of confidential scrutiny procedures by the Commission on Human Rights in 1970.
>
> (1994a, p. 10)

What Evans fails to acknowledge is that Evatt strenuously resisted attempts to override 'internal affairs' in the case of South Africa's apartheid policies, lest similar criticisms be levelled against Australia. Underlying Evatt's position was a keen appreciation that, in the future, other UN members could face hostility — especially during the 1960s — from the expanding ranks of the General Assembly's 'Third World' membership, among whom issues of race held a particular saliency. East–West issues were, therefore, only one dimension of human rights. Throughout the cold war period North–South issues gradually emerged within the UN over human rights issues.

North–South: 'cultural relativism' versus 'universalism'

By the mid-1970s, as *détente* constricted East–West conflict and the international economic recession deepened, the North–South debate became a more entrenched feature of the UN. The concept of human rights, once proclaimed to be absolute and universal, now came under increasing challenge from developing states. Åge Eknes claims that this tension between Western democracies and Third World governments arose due to their differing perceptions, not about the importance of human rights, but about what sort of rights and entitlements they should incorporate (1994, p. 100). A significant part of the debate revolved around the notion that human rights are relative and will vary in different cultures. Peter Baehr elaborates:

> ethical and moral standards differ in different places and times. These differences can only be understood against the background of the different cultural contexts these norms and values are part of. This cultural context is also assumed to determine the amount of attention that is given to human rights. There is not supposed to exist something like a universal morality, because the world has always been characterized by a plurality of cultures.

> (1994, p. 14)

There are two important points to be noted in this debate. First, although they may at first blush appear uncompromising, some modest convergence between the universalist position and that of the cultural relativists is evident. Few adherents to cultural relativism would deny that some human rights should indeed be universal. For example, the right to life, or the right to live free from torture or cruel and degrading treatment, strikes a resonance in all cultures. Similarly, those who argue for universality acknowledge that the emphasis given to human rights will be modulated according to cultural propriety and practices. 'Perhaps we sometimes fail', Luigi Bonanate admits, 'to dedicate sufficient attention to one of the main consequences of all of this: namely that the "universal" rights of citizens are planetary, whereas the corresponding "universal" rights of states are, instead, national, statist and restricted, although it is reasonable to believe that a "condensed nucleus of values and criteria universally accepted by all states" has now been developed' (1994, p. 9). Sustaining even such 'condensed' human rights will be a difficult task, especially where 'The gap between North and South in life chances and prosperity may yet prove a greater challenge to an international order based on human rights than the nuclear stand off in the Cold War' (Boyle 1995, p. 94).

Second, the debate between universalist and cultural relativist approaches to human rights suffers from political manipulation. Regimes who maintain power through repression have an interest in promoting culturally 'appropriate' human rights as a way of evading international criticism of their domestic political processes (Baehr 1994, pp. 19–20). Sydney Bailey and Sam Daws claim that this process is often played out in UN forums:

> The UN ... provides a convenient alibi, a diversion. There have been countless cases where delegates, ashamed of some national or regional misdemeanour, have used the UN to direct the world's attention elsewhere or where states guilty of internal repression or external aggression have taken an initiative at the UN in favour of some desirable goal such as the ending of colonialism or the renunciation of force.
>
> (1995, p. 104)

Moreover, those states who proclaim that human rights should be universal are often inclined to use this issue to promote other interests. Developed states who protest loudly about the abrogation of human rights do so in some cases to support their economic interests — a particular feature of recent United States–Chinese relations. Increased economic competition among states will, in all likelihood, only complicate human rights matters.

At the June 1993 World Conference on Human Rights in Vienna, disagreement between the West and the Third World over the issue of 'cultural relativism' was expressed in predominantly political economy terms, as Kevin Boyle notes: 'At its heart Asian dissidence was based on a claim that the idea of human rights was not a universal one, but a Western and developed world construct. The West, it was feared, sought to use human rights as a stalking horse, to achieve global economic dominance over the developing and poorer world' (1995, p. 84).

In essence, then, the issue of human rights needs to be seen in a broader context, moving beyond basic political rights (the core feature of the Western conceptualisations of human rights) towards economic and social rights (a key aspect of the Third World's notion of what human rights must encompass). As Johan Galtung argues, we need to become more aware and more critical of what he perceives to be the dominant Eurocentric view of human rights. In so doing, Galtung notes that we will come to see the negative role of the UN in underpinning this 'universalism': 'the notion of universal human rights essentially drawn from only one civilization with one sender only, UNGA [United Nations General Assembly], with a Security Council (Western dominated) equipped to enforce compliance, is a recipe for regional and national revolts. The similarity with the Christian Crusades (1095–1291) is only too obvious' (1994, p. 21). Perhaps Galtung goes too far. But his criticism does challenge us to think about what has become the 'accepted' view of human rights and the 'legitimacy' of this international forum for developing human rights.

Broadening and deepening human rights will pose problems for all states. Yet it is clearly in the interests of the majority of the world's population, in both the West and Third World, to ensure that human rights are pushed in this direction. The enjoyment of political freedoms, important as that might be, is simply not enough when access to economic and social resources are blocked. Conversely, the possession of economic and social rights without political rights is a template for authoritarian rule and for greater social and economic inequality. Moreover, we need to be critical of political rights that appear democratic but, in reality, merely offer the choice between identical packages. And economic rights that enhance the economic freedom of some while casting others into poverty only serve to diminish collective human rights. In the post-cold war context, '"Elections" and "free markets" may even be emerging as new ideological bases for policies destructive of human rights' (Donnelly 1992, p. 255).

The human rights agenda is thus highly political. It calls into question the political and economic legitimation of all states and cannot be confined to spotting the episodic and obvious denial of human rights somewhere 'over there'. So while the issue of human rights may sit uneasily between the domestic and international realms in terms of how jurisdictional matters are apportioned, its location serves to question the legitimacy of separating the 'domestic' and the 'international'. Some historical discussion of human rights issues during the cold war and post-cold war periods will help sharpen the analysis.

COLD WAR: HUMAN RIGHTS THEN AND NOW

For much of the cold war, issues of human rights were a key element of the East–West struggle. Providing asylum for those defecting from the Eastern bloc conferred a degree of political legitimacy on the West. The focus on human rights issues within international political forums was often designed to highlight the adverse treatment meted out to political dissidents, particularly those within the USSR. But during the cold war both sides were aligned with authoritarian regimes who systematically committed human rights abuses. Support for such regimes was often given on the basis that they were anti-communist (in the case of the USA) or pro-communist (in the case of receiving aid from the Soviet bloc and/or China). Where geostrategic advantage was considered vital, both Superpowers ignored the negation of human rights and at times protected their allies from international sanction through the use of their power of veto at the UN Security Council.

When the intensity of the cold war ebbed during the 1970s, it initially appeared that human rights issues could be separated from the bipolar competition of the 1950s and 1960s. These expectations, closely associated with the rise of multipolarity, at least in the first half of the 1970s, went largely unfulfilled. Multipolarity simply gave rise to more complex configurations of overlapping interests that went beyond the rigid bipolarity of the 1950s and 1960s. A case in point was the United States, British, and Chinese support for the Pol Pot regime in Cambodia. So, rather than hastening the progress of human rights, the initial easing in superpower tensions in fact gave rise to significant human rights abuses — in the case of Cambodia this constituted an attempt at genocide.

However, President Carter's decision in the mid-1970s to halt aid to allies of the USA on the basis of their human rights record — a consequence of the United States Congress's determination to avoid a repetition of the foreign-policy excesses of the Nixon/Kissinger period — seemed to be the harbinger of a new era of relations between states. Although criticised for being selective, Carter's initiative nonetheless rekindled earlier hopes that a human rights agenda could become a positive element in global political relations. Such optimism quickly stalled. The small window of opportunity slammed tight by the end of the 1970s as cold war antagonism reasserted itself. Human rights again featured negatively in relations between East and West. Claims and counter-claims of human rights abuses were lodged by both sides in their efforts to gain a political advantage over the other.

Unlike the period of *détente* in the 1970s, the end of the cold war in the late 1980s and early 1990s has given rise to a more sustained effort to promote human rights. For Jack Donnelly, that opportunity flows directly from the dissolution of the cold war: 'The end of the Cold War has removed the principal U.S. rationale for supporting repressive regimes, and the demise of the Soviet Union has eliminated the other major postwar pillar of support for such regimes' (1992, p. 249).

Focusing so much on the cold war and its demise has become a preoccupation for many scholars and policy-makers. To a certain extent, that focus is understandable. The historical record shows that human rights issues were used during the cold war to gain political advantage. And that both superpowers accepted significant violations of human rights as part of the 'cost' of supporting 'friendly' regimes in geostrategic locations is well documented. But as the earlier period of multipolarity in the first half of the 1970s demonstrates, there can be no guarantee that an end to the cold war will directly lead to the advancement of human rights.

It is more likely, in a world faced simultaneously with the contradictory forces of fragmentation and integration, that human rights will too quickly be discounted once again for geopolitical advantage. While the highly structured alliance systems of the cold war have altered, that old systems may be replaced by greater levels of civil disorder and conflict, with a clear potential for significant and sustained human rights abuses. The evidence from the breakup of Yugoslavia is cogent testimony to this. And moving away from that conflict, we find no shortage of other examples. Freed from the discipline of the cold war, interstate relations are likely to become far more complex. As the world moves into a situation in which the interests of states are more fluid, patterns of alignment reminiscent of pre-cold war configurations are more likely to emerge. Celebrating the demise of the cold war is thus premature, particularly when a line is drawn from the end of the cold war directly to the potential advancement of human rights. That nexus may prove to be too unrealistic. To develop this argument a little further, we need to consider human rights in more detail. There are two parts to the argument that require elaboration.

In the current post-cold war setting the immediacy of human rights abuses is evident. Whether this is manifested in the 'rape camps' of the Balkans conflict or the systematic slaughter of 'others' in Rwanda, the evidence is overwhelming and distressing. Such acts represent a clear and unequivocal abrogation of both intuitive and internationally legislated human rights. The public outrage is matched by expressions of frustration first that such things can occur in a supposedly civilised world, and then at the unwillingness of the international community to halt the offence and bring those responsible to justice. This dichotomy neatly captures the tension embedded in the human rights issue.

As we have seen in other chapters, a range of critical matters now present the international community with the inescapable question of how to solve global (transborder) problems while the dominant form of political organisation remains fixed within an interstate (bordered) system. We are, therefore, confronted by what appears to be a double-edged problem. On one level the end of the cold war has, in very real sense, liberated populations. It has also freed many states from a dependent status woven by years of bipolar politics. With this liberation of state and population, there has re-emerged the inevitable process of state-making, long suppressed by the cold war. That process, as in previous periods of state-making, is likely to be very bloody indeed. The other side of the problem is that the existing international political and military structures that have a mandate to mediate or remedy these conflicts appear deficient.

Before examining the post-cold war environment and human rights, let us first consider the role of the UN in the cold war context.

The UN during the cold war

The UN was, like any institution, created and developed within a particular context. That context, although initially not governed by a cold war logic, soon became constrained, perhaps emasculated, by the politics of superpower rivalry (Gregg 1994, p. 139). The multilateral vision that the founders of the UN had perceived would bring order to the anarchic world of competing states became subsumed beneath the blocism of the bipolar system of interlocking alliances. The UN's history is dotted with examples of it being unable to operate effectively because of the political intervention of the major powers represented at the Security Council. As Gareth Evans points out, 'In the first 45 years of its existence, between 1946 and 1990, the Security Council passed 646 resolutions — but saw vetos cast on 201 occasions, involving 241 actual negative votes' (1994d, p. 38). Since the 1970s the USA and Britain have used the power of veto more frequently than the USSR/Russian Federation (Bailey 1994, p. 127).

The view that the potential of the UN was once 'defeated' by the cold war does not necessarily imply that its recent 'liberation' will usher in a 'new world order'. The UN–Iraqi war — during which proclamations of the arrival of a 'new world order' abounded — bore remarkable similarities to the UN–Korean war, where the USA had also largely determined the parameters of the war and had done so with a clear perception of its interests. The multilateral forces that took part in both conflicts were constructed around overlapping interests rather than notions of collective or cooperative security. These two military engagements, standing, as they do, at either end of the cold war, suggest that the ability of the UN to act will remain compromised by the interests of the major powers. The UN, as with its predecessor, the League of Nations, is a political organisation created by states for states. Despite its important role in developing the human rights agenda over the past fifty years, the institution remains an integral part of an interstate system, and as such, human rights will continue to be ignored when its members consider compliance or intervention to be counter to their interests. So, while consternation at the apparently unstoppable rate of large-scale human rights abuses in the post-cold war period concentrates, with some justification, on the lack of decisive action taken by the UN, the attention needs to be more sharply focused on those states that pursue national interests under the guise of multi-lateralism (Camilleri 1989, p. 87).

The UN does have limitations, but many of these limits are those placed upon it by some member states. One of the more significant limitations is the level of financial contributions that it is able to extract from its members. K. P. Saksena notes that, 'As of 1987, the UN Centre for Human Rights had on its staff forty-eight professionals and thirty-three personnel from other categories. For the biennium 1984–85, budget appropriations amounted to $10.43 million and $14.07 million for the biennium 1986–87' (1993, p. 87). Such small amounts of funding are clearly insufficient to sustain any human rights initiatives, let alone any significant deployment of UN forces to prevent human rights abuses through peacekeeping programs. In 1992 the Secretary-General of the UN, Boutros Boutros-Ghali, outlined the parlous state of UN finances in terms of its peacekeeping operations: 'Peace-keeping operations approved at present are estimated to cost close to $3 billion in the current 12-month period, while patterns of payment are unacceptably slow. Against this, global defence expend-itures at the end of the last decade had approached $1 trillion a year, or $2 million per minute' (1992, p. 28). This situation is hardly consis-tent with the broad statements against human rights injustices made by various political leaders. And these amounts pale against the cost of the UN action in the Gulf War: $70 billion. Thus without the resources, both economic and political, the UN is rendered ineffective, and the opportunities for it to intervene remain highly constricted. And as long as the USA remains the principal financial underwriter for the UN — providing 25 per cent of its general budget and 30 per cent of peacekeeping costs — the United States Congress, as Robert Gregg notes, will be 'conspicuously reluctant [to provide funds], except when vital US interests are at stake' (1994, p. 165). It is hardly surprising, therefore, that the states with the deepest pockets exercise disproportionate leverage over the success or otherwise of UN operations. Before we look in more detail at the role of the UN with regard to human rights, a further point needs to be noted.

What has become known as the international human rights agenda is largely a construction of the post-Second World War era. Although individual rights existed before this — and there were long-standing principles in relation to war-fighting and non-combatants to be found in the 'Just War' tradition (Walzer 1984) — it only is in recent times that human rights, as such, have become part of the interstate discourse. Noteworthy and substantial pressure for the development of a human rights agenda flowed from the collapse of the 'Just War' tradition that accompanied the Holocaust of the Second World War.

The end of the Second World War, therefore, marks the point at which human rights 'really emerged as a standard subject of international relations'. A key element in that process were the Nuremberg War Crimes Trials (1945–46), 'at which leading Nazis were prosecuted under the novel charge of crimes against humanity. Many have seen this ex post facto prosecution, punishment for acts that although clearly immoral were not legally prohibited at the time they were committed … Nuremberg was an important step toward international action against human rights violations' (Donnelly 1993, p. 7). Not long after Nuremberg, however, the Allied powers who had sat in judgement on the war crimes of the Third Reich, were readily ignoring their own or their allies' human rights abuses, which were supposedly justified in the context of the developing cold war order.

Yet, paradoxically, the cold war, 'a time of pervasive anti-humanitarian interventions by both superpowers, … was also the period in which human rights first became an established subject of international relations' (Donnelly 1992, p. 250). One possible extension of that view is that the issue of human rights — having been a fundamental part of the peculiar and particular politics of the cold war — may no longer have a firm basis in any future interstate system, particularly one in transition and in which the uncertainties of post-cold war international relations dominate. A key feature of that transition phase is likely to be a new period of state dissolution (fragmentation) and state-making (integration). The working through of these processes will result, as it has throughout history, in large population shifts and growing numbers of refugees.

Human rights in the post-cold war world: the refugee crisis

In the 1950s the UN High Commission for Refugees was responsible for around one million refugees. By 1991 the global refugee population was conservatively estimated to be around fifteen million (Cunliffe 1995, p. 279). Since that time the refugee population has grown rapidly. Furthermore, that estimate does not include the migrating populations who, although not officially classified as such, are refugees within the borders of their nations. Shifting populations of refugees — whether as a result of war, civil disorder, persecution, adverse government policy, or simply the search for greater political and economic freedoms — can be regarded as the direct result of human rights abuses. Andrew Shacknove, drawing upon the connection between the end of the cold war and the growing global refugee populations, makes two key points:

> The transformation of international politics since the demise of the Soviet bloc means that the human rights community must offer a new justification for refugee protection suitable to these changed circumstances. Gradually, refugee policy may be moved further in the direction of human rights. But governments will be moved only on their own terms, primarily in the discourse of national interest. When interests of State are fundamentally at odds with other values, as is increasingly the case with asylum, then it is unlikely that compassion, solidarity or human rights will prevail.
>
> (1993, p. 518)

The rigid bipolarity of the cold war contained populations within states, particularly those on the Eurasian and Asian land mass. Freedom of movement for millions was thus restricted, certainly in terms of transborder migration, and importantly in terms of internal migration as well. The symbolism, then, of the breached Berlin Wall was not so much that it represented the final act in the collapse of communism but that populations who have historically moved across Europe and Eurasia were once again able to migrate. The political peculiarities generated during the cold war period meant that those able to leave the East were welcomed in the West as 'asylum-seekers' and not refugees. Allowing the passage of a few 'asylum seekers' carried the explicit message that Eastern bloc states were 'repressive, backward, inferior' (Shacknove 1993, p. 520).

With the re-emergence of ethno-nationalist conflict, and of state-disintegrating and state-making forces, populations of refugees have been created. With the political controls of the cold war no longer in place, these populations are once again mobile. But there is no longer any political advantage in encouraging 'asylum-seekers'. The problem for the international community, as Shacknove observes, is that it has as yet shown itself unwilling to extend the notion of human rights to include refugees. Refugee status remains bound up in narrow cold war conceptualisations of political persecution (asylum), with a human right to migrate freely not considered sufficient grounds for refugee status. Perhaps this unwillingness of states to reach agreement on how to deal with the refugee issue is more a 'reaction to the changing nationality of the refugee population, whose origins now spring primarily from "Third World" crises rather than from "Cold War" issues' (Cunliffe 1995, p. 279).

Yet the issue of shifting populations — a factor that only in recent times has become increasingly subject to restriction — is perceived to be so threatening to state sovereignty that greater and more repressive controls are now thought to be necessary (Shacknove 1993, pp. 519–20). In this instance it should be recalled that states have historically absorbed significant influxes of immigrants — many refugees — when it was considered to be in the interests of the states to do so. For example, postwar immigration policy in Australia, as part of the Chifley government's nation-building strategy, took 'selected' refugees who had been displaced by the Second World War (Harris 1993, p. 35). James Jupp and Marie Kabala note that 'Immigration policy has always been motivated by economic objectives ... [and] is modified in the face of economic realities' (1993, p. 247). In the Australian context, immigration has played a significant part in the political, social, cultural, and economic development of Australia. So too,

Australia's foreign relations, particularly those with neighbouring states, have been shaped by its immigration policy.

AUSTRALIA IN A CHANGING WORLD

The rapid recent development of Australia's regional ties has been discussed in Chapter 4, and this process has involved — from the Whitlam government through to the Keating governments — a strong and unambiguous rejection of racism, both domestically and in Australia's foreign relations. After becoming minister for foreign affairs and trade in September 1988, Gareth Evans continued with this theme and extended it through what he terms 'good international citizenship' (GIC).

Good international citizenship

Although it is difficult to determine how the GIC concept became an integral part of Evans's foreign-policy approach, it is clear that he saw GIC as important to Australia's national interest. Evans explained GIC in the following way: 'Good international citizenship is perhaps best described, not least for the cynical, as an exercise in enlightened self-interest: an expression of idealistic pragmatism. Our refugee program, for example, shows how we can be faithful to humanitarian concerns and, in the process, also acquire for Australia human resources and skills which strengthen our economy and enrich our society' (1989b, p. 13). Despite ranking below economic and military security issues, GIC was, for Evans, a device that allowed movement beyond a rigid assessment of national interests in areas where ethical and moral considerations are not generally accepted criteria in policy formulation. Yet GIC was, as the above quotation makes clear, interest-based. Evans was thus not advocating a full-blooded commitment to global community but one tempered, as he admits, by a realistic appreciation of how the world is and what Australia's foreign policy can actually achieve. In this way, Evans highlighted the rationalist underpinnings of GIC.

Both David Goldsworthy (1994, p. 6) and Andrew Linklater (1992, p. 22) draw attention to the fact that there is much in the GIC concept that resonates with the work of the Australian international-relations scholar Hedley Bull. During the 1970s, when Australia's foreign policy was attempting to adapt to the changing circumstances, Bull argued strongly for an approach that would 'allow us to contribute to human purposes beyond ourselves — to play some part in advancing the common goals, such as they are, of our region and of the world as a whole, and thus to achieve some measure of dignity and self-respect' (1973, p. 137). Sensitivity to the changing nature of international and regional political configurations is clearly evident in Bull's rationalist approach (Lawler 1992, p. 245). However, the 1970s are not the 1990s, and Evans found himself in a very different and difficult political context indeed. Some of that difference and difficulty can be attributed to the end of the cold war. New opportunities and constraints have accompanied the change in global politics. But to a greater extent, what confronted Australia was the narrowing of domestic and foreign policies. As Evans cogently points out:

> our opportunity to influence events depends, in this area more than anywhere else, on keeping our domestic house absolutely in order. Our ability to secure advances in the areas of human rights, refugees ... rests on our being, and continuing to be seen to be ... a country which articulates and applies human rights and similar principles with absolute consistency and impartiality ... Hypocrites are not merely disliked, in international relations as elsewhere. If they are our size, they are ignored.
>
> (Evans & Grant 1995, pp. 37–8)

For Andrew Linklater, there resides within the GIC discourse a strong potential to move beyond narrow conceptualisations of self-interest and a promise 'to overcome that conflict between citizenship and humanity which has been such a recurrent feature of theory and practice of international relations' (1992, p. 22). This is a very ambitious burden for GIC to bear. And it is one that Evans himself appeared not to be willing to impose. If we lower our sights from this lofty goal we can nevertheless examine the GIC concept in relation to Australia's policy towards refugees.

Australia and its foreign-policy approach to refugees

Historically, Australia's response to accepting refugees has been mixed (Harris 1993, p. 35). Although it accepted significant numbers of refugees at the end of the Second World War and into the late 1940s, it did so selectively, and from the perspective of nation-building and economic development. So too, Australia's record of granting asylum to a substantial number of Indo-Chinese refugees after the Vietnam War was designed more to pre-empt and control a potential exodus of refugees seeking to make Australia their final destination (Hardy 1992, pp. 153–4). In the case of the Indo-Chinese refugees, it is important to note two points. First, Australia's involvement in the Vietnam War and its active support for United States trade sanctions against Vietnam after the war helped create a refugee situation for which Australia must bear some responsibility (Einfeld 1993, pp. 46–8). Proclaiming one's own moral or ethical virtue for accepting Indo-Chinese refugees is thus hypocritical in the extreme. Second, a proportion of Indo-Chinese refugees arrived by boat on Australian shores. The Australian authorities at that time had few options but to accept these arrivals.

Between 1981–82 Australia resettled nearly 22 000 refugees and special humanitarian cases out of a total immigrant intake of around 120 000 persons Refugees represented a relatively small portion of Australia's overall immigrant numbers. Even so, from the early 1980s the Australian government pursued policies aimed at restricting the level of refugees as a proportion of migrant entry. However, the most significant shift in the refugee policy came in 1989. The policy was tightened in response to what was perceived to be another 'wave' of Indo-Chinese refugees arriving by boat in November 1989. Unlike those before them, these refugees were not granted automatic asylum. Moreover, these refugees were not provided with legal assistance and were held in 'detention for almost a year before the Minister, Mr Hand, acceded to requests from concerned community groups to allow lawyers to see the boat people' (Poynder 1993b, p. 289). Without access to legal assistance, the refugees were left to cope alone with the preparation of their applications for asylum.

Most of the refugees who have arrived in Australia by boat from Indo-China since 1989 have remained in detention while the government assessed their status. Keeping them in detention without due process has posed a problem for the government. Mary Crock notes:

> On 5 May [1992], some two days before the Federal Court was to hear arguments on whether the Cambodians should be released, the Federal Parliament passed the *Migration Amendment Act 1992* (Cth). The legislation passed through both Houses of Parliament in little more than one hour. It was given Royal Assent on the following day. The effect of the legislation was to insert into the *Migration Act 1958* (Cth), a new Division 4B preventing the release of the detainees, who were included in a new class of 'designated persons'
>
> (1993, p. 340)

During the parliamentary debate on the *Migration Amendment Act 1992*, it was clear that the government sought the new legislation to specifically deter 'boat people'. Minister Hand stated: 'The Government is conscious of the extraordinary nature of the measures which will be implemented by the amendment aimed at the boat people. I believe it is crucial that all persons who come to Australia without prior authorisation not be released into the community. Their release would undermine the Government's strategy for determining their refugee status or entry claims'. Furthermore, as Hand openly acknowledged, 'No law other than the Constitution will have any impact on it' (House of Representatives, *Debates*, 5 May 1992, p. 2372). For the refugees to challenge the amendment, they would have to bring their case to the High Court — an expensive process and one that would delay having their case resolved quickly. Subsequent changes to the Migration Act in September 1994 now 'confines the status of non-citizens to either lawful or unlawful non-citizens' (Palmer 1994, p. 3). For the 'boat people', their status as designated persons under the 1994 amendment means that they 'are to remain in immigration detention until they are removed from Australia or they are granted a visa' (Palmer 1994, p. 6).

In comparison with the refugee populations in other countries in which refugees seek asylum, the 500 or so refugees incarcerated at Port Hedland and Villawood in 1995 was minuscule: 'Pakistan shelters over 3 million Afghans, Thailand shelters over 300 000 Cambodians and Sudan is a refuge for over 370 000 Ethiopians' (Hardy 1992, p. 148). This highlights what is a fundamentally flawed approach towards the current refugees in Australia. It is a highly selective policy in that it is principally only the 'boat people', some 500 persons out of the 16 000 refugee applicants currently in Australia, who are kept in detention. Clearly, as Hand accepted, the policy aimed to deter other 'boat people' from attempting the voyage to Australia. Such an approach is not only misguided, cost ineffective, and ethically dubious; it is also in breach of internationally agreed human rights standards (Einfeld 1993; Chamarette 1993).

The treatment of refugees, particularly treatment as selective as that experienced by the 'boat people', has foreign-policy ramifications. Like the *Immigration Restriction Act 1901* (Cth), which as Sean Cooney points out was 'devised to implement the "White Australia Policy" and to ensure the "preservation of a British-Australian nationality"' (1994, p. 126), the current refugee internment sends unequivocal signals to Australia's neighbours and to the international community. Gareth Evans has noted, for instance, that:

> the pursuit of human rights ... need not be entirely selfless. An international reputation as a good international citizen on these issues probably can be helpful to a country in pursuing its other international interests. But more importantly, there is a real sense in which, by embracing the cause of those who have been denied their rights, we also guard and reinforce the nature of those rights themselves. The historical record shows clearly enough that rights not defended are rights easily lost.
>
> (1994c, p. 41)

While it would be easy to make the obvious point that there is a 'credibility gap' between the government's human rights rhetoric and its policies on refugees, it is important to consider the constraints that possibly prevent the government from aligning these more closely.

Clearly the Australian government, like many others, is confronted by a lack of public support for any policy that could be perceived to encourage refugees, particularly Asian 'boat people', no matter how misguided or inaccurate those perceptions may be (Lack & Templeton 1995, p. 156). It is not surprising, therefore, that many states are narrowing their definition of who is a citizen of the nation. So too, there is a process of compression in respect of who constitutes a 'legitimate' refugee (Shacknove 1985). These processes have been developed since the late 1980s to control the level of 'aliens' in many Western countries, including Australia (Birrell 1992). During a period of prolonged economic recession, the refugee 'threat' is more likely to become heightened and thus become a complex political issue with obvious electoral consequences. Apart from domestic issues, there are noteworthy foreign-policy issues involved too.

By any measure, Australia has played a strong role in facilitating the UN peace settlement in Cambodia. Accepting refugees from that country would be a *de facto* admission that the peace process was far from satisfactory. This does not, of course, legitimate the government's treatment of those refugees who remain in detention. However, it is an important constraint acting upon the government.

On a more positive note, the Keating government, on Christmas Day 1991, gave accession to the First Optional Protocol to the International Covenant on Civil and Political Rights. This instrument allows for an external appeal process for those who consider that their rights have been denied or infringed by the Australian authorities. Although the Protocol is not binding, it does represent a significant move towards ensuring that, among other rights, the right to due process is upheld (Charlesworth 1991). Legal representatives for the incarcerated refugees have lodged an appeal under the First Protocol. An acid test for the government will be its reaction to any adverse decision that may result from this evaluation process. It may well be that the government is searching for a point of exit from a policy that it recognises brings it no credit. But the history of the incarcerated refugees indicates that the government wants them to be repatriated, not released.

CONCLUSION

The term 'human rights' has been part of the vocabulary of statecraft since the end of the Second World War. The extent to which human rights can be comfortably accommodated within the realist framework reveals a tension between the notion of universal human rights and the particular and interest-bound perspectives of individual states. And within the debate about whether universal human rights should be conferred on all, there is a sharp dissonance between developed and developing countries. Human rights are considered by some developing states to be a political tool of the developed world, whose aim, it is believed, is to continue old patterns of economic and cultural dominance.

During the cold war period, refugees were often granted asylum on the basis that their arrival demonstrated a degree of political legitimacy on the part of the receiving state. In the post-cold war era it would appear that refugees are no longer the bearers of political capital. Unfortunately the changed international political environment has brought with it a resolve on the parts of many states to pursue policies to force the repatriation of so-called 'illegal' refugees. Despite the development of international covenants during the cold war on this human rights issue, many states — Australia among them — now seek to restrict available places in their immigration programs even for those refugees deemed to be 'legal'.

As the ranks of the world's refugees grow, the human rights aspects of protecting and providing resettlement for refugees are beginning to challenge the legitimacy of international political institutions — especially the UN. For instance, it is clear that adherence to UN principles on refugees has not lessened Australia's decidedly realist foreign-policy perceptions. This tension in Australia's foreign policy has been expressed in various ways. For example, the notion of good international citizenship finds poor expression in the inhumane detention of a small number of refugees who arrived on Australia's shores without permission.

Conclusion

With the collapse of cold war structures and their strategic imperatives in the late 1980s, policy-makers faced an array of global issues, but could no longer use the perceptual filters of superpower rivalry and ANZUS dependency to simplify the world. The concept of 'globalisation' is an attempt to describe the extensive transnational economic processes that are operating to create a world market. But the term can also refer to the global reach of industrialisation's environmental consequences; to a sense of a 'global village' and of common responsibility for economic welfare and human rights; to the vast expansion of media and communications technologies and industries that bring images of distant events into the home.

Traditional concerns about military security, which seemed so pressing to Australian foreign-policy-makers up until the 1970s, and which returned as fear of nuclear war in the 1980s, have eased. Nevertheless, there remain uncertainties about the continuing violent dimension of international politics — uncertainties created by rivalries among larger powers, by ethnonationalist wars, by military 'modernisation', and by the existence, still, of arsenals of weapons with the capacity for mass destruction.

Australia's belated self-confidence about military defence, based on the image of a defensible Australia in a relatively secure region, contrasts with a general sense of apprehension about a dismal economic future, crystallised by Paul Keating's 'banana republic' remark, made when he was treasurer in the mid-1980s. Anxiety about military security has been displaced by anxiety about economic security, by the fear of being left behind in the new global and regional economic race. The Australian state has sought to 'restructure' the economy by reducing protection and forcing a greater external orientation towards business. It has also pursued an energetic diplomatic campaign, regionally and internationally, for trade liberalisation. The jury may still be out on whether the neo-liberal strategy is a sufficiently purposeful approach for the Australian state to be taking in a world of more economically interventionist states.

The economic agenda has come to dominate Australia's foreign-policy concerns, but other agendas insinuate themselves into government policy in the 'borderless' world of globalisation. Concerns about environmental crisis, political repression, violent conflict, and human rights — both close at hand and far away — are subordinated in the search for trade and investment opportunities. But these concerns can still reformulate and reassert themselves, with pressure groups and, at times, wider public opinion responding to 'overseas' issues — East Timor, Somalia, or nuclear tests, for instance — and expecting the Australian government to work towards a 'better world'.

Which theoretical perspectives provide a framework for understanding the changing global politico-economic system? And which perspectives allow Australia to define its place in the world in a way that reflects the interests and aspirations of its citizens? Can a new 'economic realism' simply replace the power-politics realism of the cold war? And can Australia define itself as a hard-nosed (small-time) opportunist while the private sector chases deals with a degree of government support, and the political debate is focused on the extent of this support?

The neo-liberal argument for more open economies and freer trade makes a more general point about win–win outcomes and the extension of cooperation beyond the economic sphere. It connects economics to the wider vision of cooperation that underpins the rationalist promotion of greater world order and 'global governance' based on 'enlightened' self-interest. But 'world order' as the extension of law- and rule-based behaviour — such as arms-control agreements, environmental codes, and other international 'regimes' — takes considerably more effort than the easy assumptions of economic liberalism, with its international 'hidden hand'.

There are continuing demands that the concept of an emerging global community be incorporated into the Australian foreign-policy outlook, as pressure groups such as Amnesty International, aid and environment organisations, and women's organisations, all operating transnationally, develop an embryonic concept of global citizenship. Former Labor foreign minister Gareth Evans's idea that Australia should aspire to good international citizenship seemed to touch on this orientation towards 'humanity', while remaining primarily focused on rationalist and realist agendas. Business pressure groups campaign against environmental measures that may incur unwanted costs, maintaining that, in the calculus of national interests, their concerns are more important. This can mean that good international citizenship will develop a rhetoric for international and domestic audiences that simply evaporates at the point of policy implementation. Similarly, concerns about maintaining a residual alliance with the USA are not likely to circumscribe the *rhetoric* of arms control and disarmament policy — such as the pursuit of a 'nuclear free world' — even if on some issues this is in direct conflict with United States views; rather, they will hinder the translation of the rhetoric into a vigorous multilateral diplomacy capable of achieving such ends. When a small number of 'boat people' land on Australia's shores, commitment to universal human rights may be impeded by an older anxiety about 'floods' of unwanted immigrants. The issue becomes deeply contested, particularly since it is at 'home'

Global and regional challenges, and competing theoretical perspectives, have re-created a contemporary Australian 'identity crisis', as Australia has been thrust finally from the intellectual security blanket of its cold war ANZUS framework to face the world on the basis of its own interests, values, and understandings. Treaties are often written as though they must last forever; ANZUS, declared its signatories, was to remain in force 'indefinitely'. However, as Whitlam declared of SEATO in the 1970s, treaties become 'defunct', and ANZUS appears to have gone the same way in the 1990s. The ALP's near silence on ANZUS after the cold war — a silence that the Liberal Opposition of the time did little to fill — contrasts with its prodigious efforts to create a new conceptual discourse and institutional 'architecture' for 'our region', including the development of concepts such as regional military security, regional economic cooperation, and even an emergent regional 'community'; institutions such as the ARF and APEC; and some larger idea of regional 'destiny'. With a dose of extra-regional, middle-power multilateralism — demonstrated by its approach to peacekeeping in the UN, and to agriculture in the GATT — the ALP confronted the post-cold war world with considerable intellectual creativity and diplomatic activism.

The ALP's failure to achieve re-election in 1996 has been interpreted, in part, as a failure to link its vision of Australia's role in the region and the world to the interests and concerns of the wider Australian community. It was unable to establish a direct connection in Australian minds between the everyday desire for economic security and for a less violent world and the government's discourses, institutional architecture, and multilateral initiatives. It has been said that the former Labor prime minister Paul Keating (and his foreign minister, Gareth Evans) possessed a 'fatal abstraction' (Horne 1996). The Liberal-led coalition found it difficult to move the other way, from the concrete populism about jobs, youth unemployment, and 'mainstream values', to a distinctive view of Australia's place in the region and the world. In Opposition, the coalition seemed to regress to a sentiment of 'more ANZUS, more Europe, less region', and the ALP consequently attempted to cast itself as the only party capable of conducting a successful regional foreign policy. In government — stung by the ALP's criticism, and confronted by the reality and inevitability of the region — the Howard Liberal government set about displaying its regional credentials in its first diplomatic gestures, arguing that the coalition's only problem was Keating's attempt to undermine their chances. The Liberals continued the ALP's campaign in East Asia to have Australia accepted as an 'Asian' state in forthcoming Europe–Asia dialogue, as well as pursuing military security dialogues in regional forums. It soon became clear that regional bipartisanship was the aim of the new government, and that the 'continuity, consistency, and consensus' that the ALP proclaimed when it came to office in 1983 would also describe the approach of the Liberal government in 1996, but this time it was applied to regionalism, whereas Hawke's message was largely about the ALP's position on ANZUS.

But can 'the region' become the new cornerstone of foreign policy, rather than its most urgent preoccupation? To change the metaphor, is 'the region' the only idea left on centre stage, as the concepts of 'ANZUS' and 'middle power' move to the wings? A regional priority will focus foreign policy on that part of the world most closely connected to Australia's economic and military security, and maintain Australia as a player in regional forums. The ARF and APEC have been major initiatives, and Australia has made a significant contribution to whatever success they can claim. If Asia–Europe dialogues are to become an element of international diplomacy, leading to patterns of institutionalised cooperation, then it is clearly much better if Australia is 'in' as part of Asia, than excluded by both, which would be the only alternative. But a fundamental ambiguity regarding the boundaries of Australia's 'region' prevents the idea of 'Australia as part of the region' from becoming a coherent concept of international identity invested with a meaning that is widely shared by Australians and recognised by others. Struggling to invent an artificial single 'community' from an amorphous and diverse Asia — a community that converges with or includes Australia — may inhibit Australia in developing a foreign policy based on its domestic strengths or aspirations, which may include multiculturalism, representative democracy, lack of militarism, respect for civil rights, and institutions of industrial relations. The point is not that any of these features is unique to Australia, or uncontested within it, but rather that those features that define an Australian 'community' are present or absent in the vast East Asian area to such varying degrees that the search for community based on region runs the risk of becoming the search for vague abstractions or lowest common denominators. At the same time, the domestic basis of community is under strain in Australia, as some sectors catch the regional economic wave, while others are unable to gain a share of the benefits. A concept of regional community can only develop when Australians, rather than their government, invest it with meaning, and this is more likely to happen when regional engagement offers something to all Australians, not just those with capital and skills.

The idea of Australia as a 'middle power' has been partly discredited by its having been used in the past as a rationalisation for alliance-based policies (Australia as a 'middle-power ally'), and the more recent use of multilateral diplomacy to resist, as well as to advance, environmental agendas, as seen in the contrast between the responses to global warming and the proposed Antarctic 'world park'. The surge of UN activity after the cold war seemed to have declined by the mid-1990s. But the middle-power idea may prove to play a more significant role in locating Australia's place in the world into the next century. Premised on multilateral coalitions, it allows a regional focus, and even a regional priority, without excluding non-regional states and issues. Region provides a diplomatic focus appropriate to current economic and miliary security concerns; multilateralism provides a diplomatic method. If developed with an explicit commitment to rationalist norms and agendas, accounting for and attentive to regional concerns, the middle-power concept may offer a framework for practical collaboration based on shared interests with 'like-minded' states, wherever they are located. The concept may prove unsatisfying in that it takes the maxim that there are no 'permanent allies' seriously; it may also prove demanding when taken to imply an anti-hegemonic stance — a fundamental scepticism towards 'big powers', their ambitions, and rivalries. It may allow Australians to 'be themselves', even when their pursuit of Australian interests and ideals involves extensive cooperation with others. This kind of engagement, however, would require a sense of confidence on the part of Australia that it may not achieve until its economic future in the region is more substantially assured.

References

Albinski, H. 1977, *Australian External Policy under Labor*, Queensland University Press, St Lucia.

Altman, D. 1973, 'Internal Political Pressures on Australian Policy', in G. McCarthy (ed.), *Foreign Policy for Australia: Choices for the Seventies*, Australian Institute of Political Science and Angus & Robertson, Sydney.

Baehr, P. R. 1994, *The Role of Human Rights in Foreign Policy*, Macmillan, London.

Bailey, S. D. 1994, *The UN Security Council and Human Rights*, St Martin's Press, London.

Bailey, S. D. & Daws, S. 1995, *The United Nations: A Concise Political Guide*, 3rd edn, Barnes & Noble Books, Lanham, MD.

Ball, D. 1980, *A Suitable Piece of Real Estate: American Installations in Australia*, Hale & Iremonger, Sydney.

—— 1987, *A Base for Debate: The US Satellite Ground Station at Nurrungar*, Allen & Unwin, Sydney.

—— 1988, *Pine Gap: Australia and the US Geostationary Signals Intelligence Satellite Program*, Allen & Unwin, Sydney.

—— 1991, *Building Blocks for Regional Security: An Australian Perspective on Confidence and Security Building Measures (CSBMs) in the Asia/Pacific Region*, Strategic and Defence Studies Centre, RSPacS, ANU, Canberra.

Ball, D. & Kerr, P. 1996, *Presumptive Engagement: Australia's Asia–Pacific Security Policy in the 1990s*, Allen & Urwin, Sydney, in association with the Australian Foreign Policy Publications Program and the Department of International Relations, RSPacS, ANU, Canberra.

Beaumont, J. & Woodard, G. 1994, 'Perspectives on Australian Foreign Policy, 1993', *Australian Journal of International Affairs*, vol. 48, no. 1, pp. 97–106.

Beder, S. 1993, *The Nature of Sustainable Development*, Scribe Publications, Newham, Victoria.

Beetham, D. 1995, 'Introduction: Human Rights in the Study of Politics', *Political Studies*, vol. 43 (special issue), pp. 1–9.

Bell, C. 1988, *Dependent Ally*, Oxford University Press, Melbourne.

Bell, P. & Bell R. 1993, *Implicated: The United States in Australia*, Oxford University Press, Melbourne.

Bell, S. 1993, *Australian Manufacturing and the State*, Cambridge University Press, Melbourne.

Bergin, A. 1983, 'Pressure Groups and Australian Foreign Policy', *Dyason House Papers*, vol. 9, no. 3.

Birnie, P. 1995, 'Environmental Protection and Development', *Melbourne University Law Review*, vol. 20, no. 1, pp. 66–100.

Birrell, R. 1992, 'Problems of Immigration Control in Liberal Democracies: The Australian Experience', in G. P. Freeman & J. Jupp (eds), *Nations of*

Immigrants: Australia, the United States, and International Migration, Oxford University Press, Melbourne.

Blainey, G. 1966, *The Tyranny of Distance: How Distance Shaped Australia's History*, Sun Books, Melbourne.

Bonanate, L. 1994, *Ethics and International Politics*, Polity Press, London.

Booker, M. 1977, *The Last Domino: Aspects of Australia's Foreign Relations*, Sun Books, Melbourne.

Boulding, K. 1978, *Stable Peace*, University of Texas Press, Austin.

Boutros-Ghali, B. 1992, *An Agenda for Peace*, United Nations, New York.

Boyce, P. J. & Angel, J. R. (eds) 1992, *Diplomacy in the Marketplace: Australia in World Affairs*, vol. 7, *1981–90*, Longman Cheshire, Melbourne.

Boyle, K. 1995, 'Stock-Taking on Human Rights: The World Conference on Human Rights, Vienna 1993', *Political Studies*, vol. 43 (special issue), pp. 79–95.

Brechin, S. R. & Kempton, W. 1994, 'Global Environmentalism: A Challenge to the Postmaterialism Thesis?', *Social Science Quarterly*, vol. 75, no. 2, pp. 245–69.

Bridge, C. (ed.) 1991, *Munich to Vietnam: Australia's Relations with Britain and the United States since the 1930s*, Melbourne University Press, Melbourne.

Broinowski, A. 1992, *The Yellow Lady: Australian Impressions of Asia*, Oxford University Press, Melbourne.

Brown, G. 1994, *Australia's Security: Issues for the New Century*, Australian Defence Studies Centre, University College, Australian Defence Force Academy, Canberra.

Brundtland Report 1987. See World Commission on Environment and Development.

Bull, H. 1973, 'Options for Australia', in G. McCarthy (ed.), *Foreign Policy for Australia: Choices for the Seventies*, Angus & Robertson, Sydney.

—— 1975a, 'The Whitlam Government's Perception of our Role in the World', in B. D. Beddie (ed.), *Advance Australia — Where?*, Oxford University Press, Melbourne.

—— 1975b, 'The New Course of Australian Policy', in H. Bull (ed.), *Asia and the Western Pacific: Towards a New International Order*, Nelson, Melbourne, in assoc. with the Australian Institute of International Affairs.

—— 1977, *The Anarchical Society: A Study of Order in World Politics*, Columbia University Press, New York.

—— 1987, 'Britain and Australia in Foreign Policy', in J. D. B. Miller (ed.), *Australians and British: Social and Political Connections*, Methuen Australia, Sydney.

Burchill, S. 1994, *Australia's International Relations: Particular, Common and Universal Interests*, Australian Institute for International Affairs (Victorian Branch) and School of Australian and International Studies, Deakin University, Melbourne and Geelong.

Buttel, F. H. & Taylor, P. J. 1992, 'Environmental Sociology and Global Change: A Critical Assessment', *Society and Natural Resources*, vol. 5, pp. 211–30.

Camilleri, J. 1973, 'In Search of a Foreign Policy', *Arena*, nos 32–3, pp. 65–79.

—— 1980, *Australian–American Relations: The Web of Dependence*, Macmillan, Melbourne.

—— 1989, 'The Emerging Human Rights Agenda: Australia's Response', *Interdisciplinary Peace Research*, vol. 1, no. 1, pp. 87–116.

234 AUSTRALIA IN THE WORLD

Camilleri, J. & Falk, J. 1992, *The End of Sovereignty: The Politics of a Shrinking and Fragmenting World*, Edward Elgar, London.

Campbell, D. 1989, *The Social Basis of Australian and New Zealand Security Policy*, Peace Research Centre, RSPacS, ANU, Canberra.

Chamarette, C. 1993, 'Is Detention Justified?', in M. Crock (ed.), *Protection or Punishment? The Detention of Asylum-Seekers in Australia*, Federation Press, Sydney.

Charlesworth, H. 1991, 'Australia's Accession to the First Protocol to the International Covenant on Civil and Political Rights', *Melbourne University Law Review*, vol. 18, no. 2, pp. 428–34.

Chater, J. 1995, 'Peace Building and Preventive Diplomacy: Australian Initiatives at the United Nations', *Melbourne University Law Review*, vol. 20, no. 1, pp. 155–63.

Cheeseman, G. 1993, *The Search for Self-Reliance: Australian Defence since Vietnam*, Longman Cheshire, Melbourne.

Cheeseman, G. & Bruce, R. (eds) 1996, *Discourses of Danger and Dread Frontiers: Australian Defense and Security Thinking After the Cold War*, Allen & Unwin, Sydney, in association with the Department of International Relations, RSPacS, ANU, Canberra.

Cheeseman, G. & Kettle, S. (eds) 1990, *The New Australian Militarism: Undermining our Future Security*, Pluto Press, Sydney.

Chomsky, N. 1993, *Year 501*, South End Press, Boston.

Clark, C. 1973, 'Australia in the United Nations, in C. Clark (ed.), *Australian Foreign Policy: Towards a Reassessment*, Cassell, Melbourne.

—— 1974, 'Labor's Policy at the United Nations', *Australia's Neighbours*, no. 89, February–March, pp. 5–8.

Clark, G. 1967, *In Fear of China*, Lansdowne, Melbourne.

Commonwealth of Australia 1995, *Australia's Report to the UNCSD on the Implementation of Agenda 21*, AGPS, Canberra.

Commonwealth Record, 1983, vol. 8, no. 7.

Conca, K. 1993, 'Environmental Change and the Deep Structure of World Politics', in K. Conca & R. D. Lipschultz (eds), *The State and Social Power in Environmental Politics*, Columbia University Press, New York.

—— 1994, 'Rethinking the Ecology–Sovereignty Debate', *Millennium: Journal of International Studies*, vol. 23, no. 3, pp. 701–11.

Cooney, S. 1994, 'The Codification of Migration Policy: Excess Rules? — Part 1', *Australian Journal of Administrative Law*, vol. 1 (May), pp. 125–43.

Cooper, A. F., Higgott, R. A., & Nossal, K. R. 1993, *Relocating Middle Powers: Australia and Canada in a Changing World Order*, Melbourne University Press, Melbourne.

Cox, R. 1989, 'Middlepowermanship, Japan, and the Future World Order', *International Journal*, vol. 44, no. 4, pp. 823–62.

Crock, M. E. 1993, 'Climbing Jacob's Ladder: The High Court and the Administrative Detention of Asylum Seekers in Australia', *Sydney Law Review*, vol. 15, pp. 338–56.

Cunliffe, A. 1995, 'The Refugee Crises: A Study of the United Nations High Commission for Refugees', *Political Studies*, vol. 43, no. 2, pp. 278–90.

Dalby, S. 1992, 'Security, Modernity, Ecology: The Dilemmas of Post-Cold War Security Discourse', *Alternatives*, vol. 17, pp. 95–134.

—— 1996, 'Continent Adrift? Dissident Security Discourse and the Australian Geopolitical Imagination', *Australian Journal of International Affairs*, vol. 50, no. 1, pp. 59–75.

David, A. & Wheelwright, T. 1989, *The Third Wave: Australia and Asian Capitalism*, Left Book Club Co-operative, Sydney.

Day, D. 1991, 'Pearl Harbor to Nagasaki', in C. Bridge (ed.), *Munich to Vietnam: Australia's Relations with Britain and the United States since the 1930s*, Melbourne University Press, Melbourne.

Department of Defence 1987, *The Defence of Australia 1987*, AGPS, Canberra.

Department of Environment Sport and Territories (undated), *A Guide to the UN Framework Convention on Climate Change*, Department of the Environment, Sport and Territories, Canberra.

Department of Foreign Affairs and Trade 1995, *Media Release*, 12 May.

—— 1996, *Peace and Disarmament Newsletter*, April.

Destexhe, A. 1995, The Shortcomings of the 'New Humanitarianism', paper presented to the 50th Anniversary Conference of the United Nations, La Trobe University, Melbourne, July 1995.

Deudney, D. 1990, 'The Case Against Linking Environmental Degradation and National Security', *Millennium: Journal of International Studies*, vol. 19, no. 3, pp. 461–76.

DFAT. See Department of Foreign Affairs and Trade.

Dibb, P. (ed.) 1983, *Australia's External Relations in the 1980s: The Interaction of Economic, Political and Strategic Factors*, Croom Helm Australia, Canberra.

—— 1986, *Review of Australia's Defence Capabilities: A Report to the Minister of Defence*, AGPS, Canberra.

—— 1995, 'Towards a New Balance of Power in Asia', *Adelphi Papers*, no. 295, pp. 1–93.

Donnelly, J. 1992, 'Human Rights in the New World Order', *World Policy Journal*, vol. 9, no. 2, pp. 249–78.

—— 1993, *International Human Rights*, Westview Press, Boulder.

Doran, P. 1993, 'The Earth Summit (UNCED): Ecology as Spectacle', *Paradigms*, vol. 7, no. 1, pp. 55–65.

Eccleston, R. 1989, 'Indonesia in a Flap over Bob's Hawks', *Age*, 30 May, p. 4.

Einfeld, M. 1993, 'Detention, Justice and Compassion', in M. Crock (ed.), *Protection or Punishment? The Detention of Asylum-Seekers in Australia*, Federation Press, Sydney.

Eknes, Å. 1994, 'The United Nations and Intra-State Conflicts', in M. Heiberg (ed.), *Subduing Sovereignty: Sovereignty and the Right to Intervene*, Pinter Publishers, London.

Elliot, L. 1993, *Protecting the Antartic Environment: Australia and the Minerals Convention*, Australian Foreign Policy Publications Program, ANU, Canberra.

Emy, H. V. 1993, *Remaking Australia*, Allen & Unwin, Sydney.

Emy, H. V. & Hughes, O. E. 1991, *Australian Politics: Realities in Conflict*, 2nd edn, Macmillan, Melbourne

Evans, G. 1988, *Backgrounder*, no. 644, 7 December, p. A46.

—— 1989a, *Australia's Regional Security*, AGPS, Canberra (republished in Fry 1991).

—— 1989b, 'Australian Foreign Policy: Priorities in a Changing World', *Australian Outlook*, vol. 43, no. 2, pp. 1–15.

—— 1990, 'Foreign Policy and the Environment', *Australian Foreign Affairs Record*, vol. 61, no. 3, pp. 112–18.

—— 1993, *Cooperating for Peace: The Global Agenda for the 1990s and Beyond*, Allen & Unwin, Sydney.

—— 1994a, 'Cooperative Security and Intra-State Conflict', *Foreign Policy*, vol. 96, pp. 3–20.

—— 1994b, The Labor Tradition in Australian Foreign Policy, keynote address by Senator Gareth Evans, Minister for Foreign Affairs, to The Labor Tradition and Australian Foreign Policy Symposium, 5 December, ANU, Canberra.

—— 1994c, Address by Senator Gareth Evans QC, Minister for Foreign Affairs, at Award of Human Rights Law Prize, Melbourne, 8 September 1994, reprinted in Department of Foreign Affairs and Trade, *Australia and the United Nations*, AGPS, Canberra, pp. 38–46.

—— 1994d, 'The World after the Cold War — Community and Cooperation: An Australian View', *The Round Table*, no. 329, pp. 33–9.

—— 1995, Australia in East Asia and the Asia Pacific: Beyond The Looking Glass, Fourteenth Asia Lecture, by Senator Gareth Evans, Minister for Foreign Affairs, to Asia–Australia Institute, Sydney, 20 March 1995.

Evans, G. & Grant, B. 1995, *Australia's Foreign Relations: In the World of the 1990s*, 2nd edn, Melbourne University Press, Melbourne.

Evatt, H. V. 1945, *Foreign Policy of Australia: Speeches by the Rt Hon. H. V. Evatt*, M. P., Angus & Robertson, Sydney.

Falk, R. 1987, 'The Global Promise of Social Movements: Explorations at the Edge of Time', *Alternatives*, vol. 11, no. 2, pp. 173–96.

Faulkner, J. 1994, Speech by Senator the Honourable John Faulkner, Australian Minister for the Environment, Greenhouse 94 Conference, Wellington, New Zealand, 10 October 1994.

—— 1995, 'Senator John Faulkner, Opening Statement, Press Conference for the Release of Greenhouse 21C, The Federal Government's Additional Package of Greenhouse Response Measures', Parliament House, Canberra, 29 March 1995, *Media Release*.

Foreign Relations of the United States 1951 1977, US Government Printing Office, Washington DC.

Foster, J. B. 1985, 'Sources of Instability in the US Political Economy and Empire', *Science and Society*, vol. 49, pp. 167–93.

Fry, G. (ed.) 1991, *Australia's Regional Security*, Allen & Unwin, Sydney.

Galtung, J. 1994, *Human Rights in Another Key*, Polity Press, London.

Garnaut, R. 1989, *Australia and the Northeast Asian Ascendancy: Report to the Prime Minister and Minister for Foreign Affairs and Trade*, AGPS, Canberra.

Girling, J. L. S. 1977, 'Australia and Southeast Asia in the Global Balance: A Critique of the "Fraser Doctrine"', *Australian Outlook*, vol. 31, no. 1, pp. 3–15.

Gleick, P. 1989, 'The Implications of Global Climatic Changes for International Security', *Climatic Change*, vol. 15, pp. 309–25.

Goldsworthy, D. 1975, 'The Whitlam Government's African Policy', *Dyason House Papers*, vol. 1, no. 3.

—— 1994, *Being, and Being Seen to Be: Australia and Good International Citizenship*, Working Paper No. 11, Centre for International Relations, Department of Politics, Monash University, Melbourne.

Goot, M. & Tiffen, R. (eds) 1992, *Australia's Gulf War*, Melbourne University Press, Melbourne.

Graf, W. D. 1992, 'Sustainable Ideologies and Interests: Beyond Brundtland', *Third World Quarterly*, vol. 13, no. 3, pp. 553–9.

Grant, B. 1972, *The Crisis of Loyalty*, Angus & Robertson, Sydney.

Grattan, M. 1995, 'Countdown to Crisis', *Age*, 8 July, p. 17.

Gregg, R. W. 1994, *About Face? The United States and the United Nations*, Lynne Rienner Publishers, Boulder.

Hague Declaration 1989, 11 March.

Hanisch, T. 1992, 'The Rio Climate Convention: Real Solutions or Political Rhetoric?', *Security Dialogue*, vol. 23, no. 4, pp. 63–73.

Hardy, L. 1992, 'Running the Gamut: Australia's Refugee Policy', in P. Keal (ed.), *Ethics and Foreign Policy*, Allen & Unwin, Sydney, in assoc. with Department of International Relations, RSPacS, ANU, Canberra.

Harper, N. 1976, 'The American Alliance in the 1970s', in J. A. C. Mackie (ed.), *Australia in the New World Order: Foreign Policy in the 1970s*, Nelson, in association with the Australian Institute for International Affairs, Melbourne.

Harries, O. 1989, *Strategy and the Southwest Pacific: An Australian Perspective*, Pacific Security Research Institute, Sydney.

Harris, S. 1993, 'Immigration and Australian Foreign Policy', in J. Jupp & M. Kabala (eds), *The Politics of Australian Immigration*, AGPS, Canberra.

Harry, R. L. 1982, 'Ethnic Minorities in Australia and Foreign Policy', *World Review*, vol. 21, no. 1, pp. 57–74.

Harwell, M. A. 1984, *Nuclear Winter: The Human and Environmental Consequences of Nuclear War*, Springer-Verlag, New York.

Hastings, P. 1977, 'Foreign Policy Problems on our Doorsteps — Realism or Rhetoric?', *Dyason House Papers*, vol. 3, no. 4, pp. 1–4.

Hawke, R. J. L. 1989, *Our Country Our Future*, AGPS, Canberra.

Head, B. (ed.) 1983, *State and Economy in Australia*, Oxford University Press, Melbourne.

Higgott, R. 1981, 'Australia and Africa 1970–80: A Decade of Change and Growth', in C. Legum (ed.), *Africa Contemporary Record*, vol. 14, Africana, New York.

—— 1991, 'International Relations in Australia: An Agenda for the 1990s', in R. Higgott & J. L. Richardson (eds), *International Relations: Global and Australian Perspectives on an Evolving Discipline*, Department of International Relations, RSPacS, ANU, Canberra.

Higgott, R., Leaver, R., & Ravenhill, J. (eds) 1993, *Pacific Economic Relations in the 1990s*, Allen & Unwin, Sydney.

Higgott, R. & Richardson, J. L. (eds) 1991, *International Relations: Global and Australian Perspectives on an Evolving Discipline*, Department of International Relations, RSPacS, ANU, Canberra.

Holbraad, C. 1984, *Middle Powers in International Politics*, Macmillan, London.

Horne, D. 1996, 'Paul Keating's Fatal Abstraction', *Weekend Australian*, 9–10 March, p. 21.

House of Representatives, *Debates* (various volumes).

Howson, P. 1984, *The Life of Politics: The Howson Diaries*, Viking, Melbourne.

Hudson, W. J. 1992, 'The Australian People and Foreign Policy', in F. A. Mediansky (ed.), *Australia in a Changing World: New Foreign Policy Directions*, Maxwell Macmillan, Sydney.

Hurrell, A. 1994, 'A Crisis of Ecological Viability? Global Environmental Change and the Nation State', *Political Studies*, vol. 42, pp. 146–65.

Indyk, M. 1985, 'The Australian Study of International Relations', in D. Aitken (ed.), *Surveys of Australian Political Science*, Allen & Unwin, Sydney.

Ingram, D. 1979, 'Lusaka 1979: A Significant Commonwealth Meeting, *The Round Table*, no. 275, pp. 275–83.

Intergovernmental Panel on Climate Change 1990, *Climate Change: The IPCC Impacts Assessment*, AGPS, Canberra.

IPCC. See Intergovernmental Panel on Climate Change.

Jacob, M. 1994, 'Toward a Methodological Critique of Sustainable Development', *The Journal of Developing Areas*, vol. 28, no. 2, pp. 237–52.

Jenkins D. 1989a, 'The Politics of Bomber Beazley', *Sydney Morning Herald*, 24 June.

—— 1989b, 'The State within a State that Dominates Diplomacy', Sydney Morning Herald, 26 June.

Joint Committee on Foreign Affairs and Defence 1981, *Threats to Australia's Security — Their Nature and Probability*, AGPS, Canberra.

Jupp, J. 1993, 'Perspectives on the Politics of Immigration', in J. Jupp & M. Kabala (eds), *The Politics of Australian Immigration*, AGPS, Canberra.

Jupp, J. & Kabala, M. (eds) 1993, *The Politics of Australian Immigration*, AGPS, Canberra.

Karns, M. P. & Mingst, K. A. (eds) 1990, *The United States and Multilateral Institutions*, Unwin Hyman, Boston.

Keal, P. (ed.) 1992, *Ethics and Foreign Policy*, Allen & Unwin, Sydney.

Keesings 1991, *Keesings Record of World Events*, vol. 37, no. 2, p. 39 786.

Kelly, R. 1992, The UN Conference on Environment and Development (UNCED): Where to Now and What Does it Mean for Australia?, Address to the National Press Club, Canberra, 19 June 1992.

Kempton, W. 1991, 'Lay Perspectives on Global Climate Change', *Global Environmental Change: Human and Policy Dimension*, vol. 1, pp. 183–208.

Keohane, R. O. 1990, 'Multilateralism: An Agenda for Research', *International Journal*, vol. 45, Autumn, pp. 731–64.

Keohane, R. O. & Nye, J. S. 1985, 'Two Cheers for Multilateralism', *Foreign Policy*, vol. 60, Fall, pp. 148–67.

Lack, J. & Templeton, J. 1995, *Bold Experiment: A Documentary History of Australian Immigration since 1945*, Oxford University Press, Melbourne.

Lawler, P. 1992, 'The Good Citizen Australia?', *Asian Studies Review*, vol. 16, no. 2, pp. 241–50.

Lawson, S. 1995, *The New Agenda for Global Security: Cooperating for Peace and Beyond*, Allen & Unwin, Sydney, in assoc. with the Department of International Relations, ANU, Canberra.

Leaver, R. 1990, 'The Garnaut Report: The Quality of Realism', *Australian Journal of International Affairs*, vol. 44, no. 1, pp. 21–8.

—— 1993, 'Sharing the Burdens of Victory: Principles and Problems of a Concert of Powers', in R. Leaver & J. L. Richardson (eds), *The Post-Cold War Order: Diagnoses and Prognoses*, Allen & Unwin, Sydney, in assoc. with the Department of International Relations, RSPacS, ANU, Canberra .

Leaver, R. & Richardson, J. L. (eds) 1993, *The Post-Cold War Order: Diagnoses and Prognoses*, Allen & Unwin, Sydney, in assoc. with the Department of International Relations, RSPacS, ANU, Canberra.

Leggett, J. & Hohnen, P. 1992, 'The Climate Convention: A Perspective from the Environmental Lobby', *Security Dialogue*, vol. 23, no. 4, pp. 75–81.

Linklater, A. 1992, 'What is a Good International Citizen?', in P. Keal (ed.), *Ethics and Foreign Policy*, Allen & Unwin, Sydney, in assoc. with the Department of International Relations, RSPacS, ANU, Canberra.

Lonie, J. 1971, 'The Dunstan Government', *Arena*, no. 25, pp. 57–73.

Lowe, D. 1994, Divining a Labor Line: Conservative Constructions of Labor's Foreign Policy, 1944–49, paper presented to The Labor Tradition in Australian Foreign Policy conference, December 1944, ANU, Canberra.

Lowe, I. 1994, 'The Greenhouse Effect and the Politics of Long-Term Issues', in S. Bell & B. Head, *State, Economy and Public Policy in Australia*, Oxford University Press, Melbourne.

Mack, A. 1986, 'The Political Economy of Global Decline: America in the 1980s', *Australian Outlook*, vol. 40, no. 1, pp. 11–20.

—— 1988, *Nuclear Allergy: New Zealand's Anti-Nuclear Stance and the South Pacific Nuclear-Free Zone*, Peace Research Centre, RSPacS, ANU, Canberra.

—— 1989, 'Australian Defense Policy and the ANZUS Alliance', in J. Ravenhill (ed.), *No Longer an American Lake?*, Allen & Unwin, Sydney.

Mack, A. & Ravenhill, J. R. (eds) 1994, *Pacific Cooperation: Building Economic and Security Regimes in the Asia-Pacific*, Allen & Unwin, Sydney, in association with the Program on International Economics and Politics, East–West Centre, Hawaii, and the Department of International Relations, RSPacS, ANU, Canberra.

Mackie, J. A. C. 1976, 'Australian Foreign Policy: From Whitlam to Fraser', *Dyason House Papers*, vol. 3, no. 1, pp. 1–5.

MacMillan, J. & Linklater, A. (eds) 1995, *Boundaries in Question: New Directions in International Relations*, Pinter, London.

MacNeill, J. 1989–90, 'The Greening of International Relations', *International Journal*, vol. 45, pp. 1–35.

McQueen, H. 1991, *Japan to the Rescue: Australian Security around the Indonesian Archipelago during the American Century*, Heinemann, Melbourne.

Maher, M. 1992, 'The Media and Foreign Policy', in P. J. Boyce & J. R. Angel (eds) 1992, *Diplomacy in the Marketplace: Australia in World Affairs*, vol. 7, 1981–90, Longman Cheshire, Melbourne.

Malik, J. M. 1992, *The Gulf War: Australia's Role and Asian-Pacific Responses*, Strategic and Defence Studies Centre, RSPacS, ANU, Canberra.

Martell, L. 1994, *Ecology and Society*, Polity Press, Cambridge.

Matthews, T. & Ravenhill, J. 1988, 'Bipartisanship in the Australian Foreign Policy Elite', *Australian Outlook*, vol. 42, no. 1, pp. 9–20.

—— 1993, Strategic Trade Policy: The East Asian Experience, paper presented to the Annual Conference of the Australian Political Studies Association, Monash University, Melbourne, 29 September–10 October, p. 35.

Meaney, N. 1991, 'Expressing and Examining the Confusions and Complexities of a Transitional Era', *Canberra Bulletin of Public Administration*, no. 68, March, pp. 158–61.

Mediansky, F. A. 1972, 'Now Here is our Foreign Policy', *Current Affairs Bulletin*, vol. 49, no. 4, pp. 99–112.

—— 1992, 'The Development of Australian Foreign Policy', in P. J. Boyce & J. R. Angel (eds), *Diplomacy in the Marketplace: Australia in World Affairs*, vol. 7, 1981–90, Longman Cheshire, Melbourne.

Menzies, R. G. 1967, *Afternoon Light: Some Memories of Men and Events*, Cassell, Melbourne.

Millar, T. B. 1978, *Australia in Peace and War: External Relations Since 1788*, Australian National University Press, Canberra.

Miller, J. D. B. 1971, 'Notes on Australian Relations with South Africa', *Australian Outlook*, vol. 25, no. 2, pp. 132–40.

—— 1974a, 'Australian Foreign Policy: Constraints and Opportunities — I', *International Affairs*, vol. 50, no. 2, pp. 229–41.

—— 1974b, 'Australian Foreign Policy: Constraints and Opportunities — II', *International Affairs*, vol. 50, no. 3, pp. 425–38.

Mische, P. M. 1989, 'Ecological Security and the Need to Reconceptualize Sovereignty', *Alternatives*, vol. 14, no. 4, pp. 389–427.

Morgenthau, H. 1960, *Politics among Nations: The Struggle for Power and Peace*, 3rd edn, Knopf, New York.

Neuhaus, M. E. K. 1988, 'A Useful CHOGM: Lusaka 1979', *Australian Outlook*, vol. 42, no. 3, pp. 161–5.

Nilsson, S. & Pitt, D. 1994, *Protecting the Atmosphere: The Climate Change Convention and its Context*, Earthscan Publications, London.

Norse, D. 1994, 'Multiple Threats to Regional Food Production: Environment, Economy, Population?', *Food Policy*, vol. 19, no. 2, pp. 133–48.

Nossal, K. 1993, 'Middle Power Diplomacy in the Changing Asia-Pacific Order: Australia and Canada Compared', in Leaver, R. & Richardson, J. L. (eds), *The Post-Cold War Order: Diagnoses and Prognoses*, Allen & Unwin, Sydney.

Nuclear Proliferation News.

Nye, R. 1990, 'Soft Power', *Foreign Policy*, no. 80, pp. 153–71.

O'Neill, N. F. K. & Handley, R. 1994, *Retreat from Injustice: Human Rights in Australian Law*, Federation Press, Sydney.

Palfreeman, A. C. 1988, 'The Political Objectives', in F. A. Mediansky & A. C. Palfreeman (eds), *In Pursuit of National Interests: Australian Foreign Policy in the 1990s*, Permagon Press, Sydney.

Palmer, A. 1994, Calling Australia Home? The Mandatory Detention of Unauthorised Boat Arrivals, Human Rights Law Prize winning paper, unpublished, Melbourne University.

Patterson, M. 1993, 'The Politics of Climate Change after UNCED', *Environmental Politics*, vol. 2, no. 4, pp. 174–90.

—— 1995, 'Radicalizing Regimes? Ecology and the Critique of IR Theory', in J. MacMillan & A. Linklater (eds), *Boundaries in Question: New Directions in International Relations*, Pinter, London, pp. 212–26.

Pemberton, G. 1994, The Labor Tradition in Australian Foreign Policy, paper presented to The Labor Tradition in Australian Foreign Policy Conference, Department of Political Science, ANU, Canberra, December 1994.

Pettman, J. 1992, 'National Identity and Security', in G. Smith & St. J. Kettle (eds), *Threats without Enemies: Rethinking Australia's Security*, Pluto Press, Sydney.

Phillips, D. H. 1988, *Ambivalent Allies: Myth and Reality in the Australian–American Relationship*, Penguin, Melbourne.

Polakas, J. 1980, 'Economic Sanctions: An Effective Alternative to Military Coercion', *Brooklyn Journal of International Law*, vol. 6, no. 2, pp. 289–320.

Pons, X. 1994, *A Sheltered Land*, Allen & Unwin, Sydney.

Poynder, N. 1993a, 'Human Rights Law and the Detention of Asylum-Seekers', in M. Crock (ed.), *Protection or Punishment? The Detention of Asylum-Seekers in Australia*, Federation Press, Sydney.

—— 1993b, 'Marooned in Port Hedland', *Alternative Law Journal*, vol. 18, no. 6, pp. 272–89.

Ravenhill, J. (ed.) 1989, *No Longer an American Lake?*, Allen & Unwin, Sydney.

Ravenhill, J. 1991, 'International Political Economy: An Australian Perspective', in R. Higgott & J. L. Richardson (eds), *International Relations: Global and Australian Perspectives on an Evolving Discipline*, Department of International Relations, RSPacS, ANU, Canberra.

Renouf, A. 1979, *The Frightened Country*, Macmillan, Melbourne.

—— 1983, *Let Justice Be Done: The Foreign Policy of Dr H. V. Evatt*, University of Queensland Press, St. Lucia.

Richardson, J. L. 1991, 'Debates and Options for Australia', in C. Bell (ed.), *Agenda for the Nineties; Australian Choices in Foreign and Defence Policy*, Longman Cheshire, Melbourne.

Richelson, J. T. & Ball, D. 1985, *The Ties that Bind: Intelligence Cooperation between the UKUSA Countries*, Allen & Unwin, Boston.

Riddell, T. 1989, 'The Inflationary Impact of the Vietnam War', *Vietnam Generation*, vol. 1, no. 1, Winter, pp. 44–60.

Robinson, N. 1993, *Agenda 21: Earth's Action Plan*, IUCN Environmental Policy & Law Paper No. 27, Oceana Publications, New York.

Rosenau, J. 1993, 'Environmental Challenges in a Turbulent World', in K. Conca & R. D. Lipschultz (eds), *The State and Social Power in Environmental Politics*, Columbia University Press, New York.

Rowlands, I. 1991, 'The Security Challenges of Global Environmental Change', *The Washington Quarterly*, vol. 14, no. 1, pp. 99–114.

Ruggie, J. G. (ed.) 1993, *Multilateralism Matters: The Theory and Praxis of an International Forum*, Columbia University Press, New York.

—— 1994, 'Third Try at World Order? America and Multilateralism after the Cold War', *Political Science Quarterly*, vol. 109, no. 4, pp. 553–70.

Russell, E. 1994, *The International Environmental Crisis: Australia's Response*, Working Paper No. 6, Centre for International Relations, Department of Politics, Monash University, Melbourne.

Saksena, K. P. 1993, *Reforming the United Nations: The Challenge of Relevance*, Sage Publications, Newbury Park, Calif.

Sanders, J. W. 1990, 'Global Ecology and World Economy: Collision Course or Sustainable Future', *Bulletin of Peace Proposals*, vol. 21, no. 4, pp. 395–401.

Seager, J. 1993, *Earth Follies: Coming to Terms with the Global Environmental Crisis*, Routledge, New York.

Shacknove, A. 1985, 'Who is a Refugee?', *Ethics*, vol. 95, no. 2, pp. 274–84.

—— 1993, 'From Asylum to Containment', *International Journal of Refugee Law*, vol. 5, no. 4, pp. 516–33.

Singer, F. 1992, 'Warming Theories Need Warning Label', *The Bulletin of the Atomic Scientists*, vol. 48, no. 5, pp. 34–9.

Sinner, J. 1994, 'Trade and the Environment: Efficiency, Equity and Sovereignty Considerations', *Australian Journal of Agricultural Economics*, vol. 38, no. 2, pp. 171–87.

Smith, G. 1986, 'Dibb on the Defensive', *Peace Magazine Australia*, no. 1, August–September, pp. 25–35.

—— 1992a, 'Demilitarising Security', in G. Smith & St J. Kettle (eds), *Threats without Enemies: Rethinking Australia's Security*, Pluto Press, Sydney.

—— 1992b, 'The State and the Armed Forces: Defence as Militarism', in M. Muetzelfeldt (ed.), *Society, State and Politics in Australia*, Pluto Press, Sydney, pp. 326–59.

Smith, H. 1992, 'Internal Politics and Foreign Policy', in F. A. Mediansky (ed.), *Australia in a Changing World: New Foreign Policy Directions*, Maxwell Macmillan, Sydney.

Smith, H. 1994 (ed.), *International Peacekeeping: Building on the Cambodian Experience*, Australian Defence Studies Centre, Canberra.

Solis Trejo, M. T. 1994, Australia and Japan in the South Pacific Island Countries: A General Appraisal through the Concepts of Structural Power, unpublished thesis, Deakin University, Geelong.

Spender, P. 1969, *Exercises in Diplomacy*, Sydney University Press, Sydney.

Stilwell, R. 1993, *Economic Inequality*, Pluto Press, Sydney.

Strange, S. 1986, *Casino Capitalism*, Blackwell, Oxford.

—— 1988, *States and Markets: An Introduction to International Political Economy*, Pinter, London.

—— 1991, 'New World Order: Conflict and Cooperation', *Marxism Today*, January.

Sullivan, D. 1995, 'The Poverty of Australian Defence Studies: The "Secure Australia Project" and its Critics', *Australian Journal of Political Science*, vol. 30, no. 3, pp. 146–57.

Teichmann, M. 1966, *Australia: Armed and Neutral?*, Victorian Fabian Society, Melbourne.

Thurow, L. 1993, *Head to Head*, Allen & Unwin, New York.

Toohey, B. & Wilkinson, M. 1987, *The Book of Leaks*, Angus & Robertson, Sydney.

United Nations Conference on Environment and Development 1992, *Agenda 21: Programme of Action for Sustainable Development*, United Nations, New York.

van Boven, T. 1991, 'Prevention of Human Rights Violations', in A. Eide & J. Helgesen (eds), *The Future of Human Rights Protection in a Changing World*, Norwegian University Press, Oslo.

Vincent, R. J. 1986, *Human Rights and International Relations*, Cambridge University Press, Cambridge.

Visvanathan, S. 1991, 'Mrs. Brundtland's Disenchanted Cosmos', *Alternatives*, vol. 16, pp. 377–84.

Walker, R. B. J. 1988, *One World, Many Worlds: Struggles for a Just World Peace*, Lynne Rienner Publishers, Boulder.

Walsh, J. & Munster, G. 1980, *Documents on Australian Defence and Foreign Policy, 1968–1975*, self-published, Hong Kong.

Walzer, M. 1984, *Just and Unjust Wars: A Moral Argument with Historical Illustrations*, Pelican Books, Harmondsworth.

Ward, B. & Dubos, R. 1972, *Only One Earth: The Care and Maintenance of a Small Planet*, André Deutsch, London.

WCED. See World Commission on Environment and Development.

Whitlam, E. G. 1973, *Australian Foreign Policy: New Directions, New Definitions*, Australian Institute of International Affairs, Brisbane.

—— 1985, *The Whitlam Government, 1972–1975*, Viking, Melbourne.

Wight, M. 1978, 'Is the Commonwealth a non-Hobbesian Institution?', *Journal of Commonwealth and Comparative Politics*, vol. 16, no. 2, pp. 119–34.

Wilkes, J. (ed.) 1967, *Communism in Asia: A Threat to Australia?*, Angus & Robertson, Sydney.

Willesee, D. (1975), 'Australian Foreign Policy for the 1970s', in B. D. Beddie (ed.), *Advance Australia — Where?*, Oxford University Press, Oxford.

World Commission on Environment and Development 1987, *Our Common Future*, Oxford University Press, Oxford.

Acknowledgments

The authors and publisher are grateful to the following copyright holders for granting permission to reproduce various extracts in this book:

Adelphi Papers and P. Dibb for extract from 'Towards a New Balance of Power in Asia'.

AGPS for extracts from Australia, House of Representatives, *Debates*, various volumes; P. Dibb, *Review of Australia's Defence Capabilities: A Report to the Minister of Defence*; Department of Defence, *The Defence of Australia 1987*; R. J. Hawke, *Our Country Our Future*; and Address by Senator Gareth Evans QC, Minister for Foreign Affairs, at Award of Human Rights Law Prize, Melbourne, 8 September 1994, reprinted in Department of Foreign Affairs and Trade, *Australia and the United Nations*.

Allen & Unwin (Australia) for extracts from R. Leaver & J. L. Richardson (eds), *The Post-Cold War Order: Diagnoses and Prognoses*; G. Fry (ed.), *Australia's Regional Security*; and H. V. Emy, *Remaking Australia*, Allen & Unwin, Sydney.

Allen & Unwin (New York) for extract from L. Thurow, *Head to Head*.

Australian Journal of Agricultural Economics and J. Sinner for extract from 'Trade and the Environment: Efficiency, Equity and Sovereignty Considerations'.

Australian Institute for International Affairs (Victorian Branch) for extract from S. Burchill, *Australia's International Relations: Particular, Common and Universal Interests*.

Barnes & Noble Books for extract from S. D. Bailey & S. Daws, *The United Nations: A Concise Political Guide*, 3rd edn.

Cambridge University Press (UK) for extract from R. J. Vincent, *Human Rights and International Relations*.

Climatic Change and P. Gleick for extract from 'The Implications of Global Climatic Changes for International Security'.

Department of Environment, Sport and Territories for extract from *A Guide to the UN Framework Convention on Climate Change*.

G. Evans for extracts from 'The Labor Tradition in Australian Foreign Policy'; and 'Australia in East Asia and the Asia Pacific: Beyond The Looking Glass', Fourteenth Asia Lecture.

Federation Press for extract from N. F. K. O'Neill & R. Handley, *Retreat from Injustice: Human Rights in Australian Law*.

Foreign Policy and G. Evans for extract from 'Cooperative Security and Intra-State Conflict'.

Interdisciplinary Peace Research and J. Camilleri for extract from 'The Emerging Human Rights Agenda: Australia's Response'.

International Journal and R. Cox for extracts from 'Middlepowermanship, Japan, and the Future World Order'; and J. MacNeill, 'The Greening of International Relations'.

International Journal of Refugee Law and A. Shacknove for extract from 'From Asylum to Containment'.

R. Kelly for extract from 'The UN Conference on Environment and Development (UNCED): Where to Now and What Does it Mean for Australia?'.

Macmillan (UK) for extract from P. R. Baehr, *The Role of Human Rights in Foreign Policy*.

Marxism Today and S. Strange for extract from 'New World Order: Conflict and Cooperation'.

Melbourne University Press for extracts from G. Evans & B. Grant, *Australia's Foreign Relations: In the World of the 1990s*.

Paradigms and P. Doran for extract from 'The Earth Summit (UNCED) Ecology as Spectacle'.

Pluto Press for extracts from M. Muetzelfeldt (ed.), *Society, State and Politics in Australia*; and G. Cheeseman & S. Kettle (eds), *The New Australian Militarism: Undermining our Future Security*.

Political Studies and D. Beetham for extract from 'Introduction: Human Rights in the Study of Politics'.

The Round Table and G. Evans for extract from 'The World after the Cold War — Community and Cooperation: An Australian View'.

Sydney Law Review and M. Crock for extract from 'Climbing Jacob's Ladder: The High Court and the Administrative Detention of Asylum Seekers in Australia'.

United Nations for extracts from B. Boutros-Ghali, *An Agenda for Peace*; and Chapter 5 of the UN Charter, 'International Economic and Social Co-operation'.

Index